Computational Triangulation in Mathematics Teacher Education

Advancements in Learning and Instruction

More information about this series can be found at
https://novapublishers.com/product-category/series/advancements-in-learning-and-instruction/

Education in a Competitive and Globalizing World

More information about this series can be found at
https://novapublishers.com/product-category/series/education-in-a-competitive-and-globalizing-world/

Sergei Abramovich

Computational Triangulation in Mathematics Teacher Education

DOI: https://doi.org/10.52305/IBJL8877

Library of Congress Cataloging-in-Publication Data

ISBN: 979-8-89530-969-8 (Softcover)
ISBN: 979-8-90134-016-5 (eBook)

Published by Nova Science Publishers, Inc. † New York

Contents

Preface

The term "computational triangulation" grew out of the author's experience using computers with future K-12 teachers of mathematics of the United States and Canada. The university where the author works is in Upstate New York on the border with Ontario province of Canada, and many students enrolled in educational programs are Canadians pursuing their master's degrees in the teaching field. "Computational triangulation" is a new term apparently not used in the literature, either mathematical or educational, before 2023. Linguistically, the term was formed when the author was invited to contribute a paper to the special issue "*Computational social science and complex systems" of MDPI Computation* (Abramovich, 2023). The term also stems from the author's interest in learning how social sciences research was enriched by the triangulation approach to sociology. The use of sophisticated and diverse software tools by teacher candidates' learning mathematics, both as a subject matter and a teaching discipline, allows them to appreciate the role of the intrinsic complexity of the tools in the development of deep mathematical knowledge required nowadays for teaching mathematics across the grades (Conference Board of Mathematical Sciences, 2012, p. 31; Association of Mathematics Teacher Educators, 2017, p. 46; Ontario Ministry of Education, 2020, p. 92). The book provides examples of using digital tools in support of computational triangulation in mathematics education as multiple ways of solving a problem. The examples differ by grade level, computational difficulty, and conceptual complexity.

Whatever the meaning of the word "triangulation", it has mathematical origin going back to antiquity as a technique of indirect measurement by constructing triangles. Much later, the idea of validation of mathematical propositions and simplification of their proofs using alternative reasoning strategies had become a commonplace of pure and applied mathematical research. In the 21st century, the pedagogical idea of encouraging teachers of mathematics to act as mathematicians when working with their students (Conference Board of the Mathematical Sciences, 2012) and the availability of different computational tools in the modern-day classroom (National Council of Teachers of Mathematics, 2014) suggest using the notion of computational triangulation in mathematical preparation of schoolteachers. The book covers topics across the entire school mathematics and beyond including arithmetic, algebra, geometry, trigonometry, probability, number theory, and discrete mathematics. Historical aspects and cultural origins of many mathematical ideas are discussed as appropriate. Given the international makeup of the author's students, the book connects the notion of computational triangulation to (available in English) mathematical standards and recommendations for teachers worldwide including Australia, Canada, Chile, England, Japan, Korea, Singapore, South Africa, and the United States. Included in the book solicited reflections by teacher candidates (the author's students) indicate their readiness to implement the pedagogy of the triangulated perspective on computational problem solving and problem posing in their own classrooms.

The book demonstrates the applicability of the notion of triangulation, typically used in social sciences research, to computationally enhanced mathematics education, in general, and to the preparation of future K-12 teachers of the subject matter, in particular. In the age of technology, with the ubiquity of computational thinking in diverse disciplines, the methodology of triangulation became integrated with advances in digital technology, allowing for scientific experiments to be validated by more than one computational instrument, thus enhancing the credibility of problem solving. The book provides mathematics education examples of using digital tools in support of computational triangulation as the social sciences construct including the triangulation *within* and *between* methods (McFee, 1992) and the triangulation *by data sources* (Denzin, 1970). It may be of interest to faculty in mathematics and education departments of colleges and universities who are assigned teaching courses for future teachers of the subject matter. The book can support graduate/doctoral level mathematics education programs offering studies of problem solving in the information age. Finally, the book may motivate using the ideas of computational triangulation within the variety of tertiary courses in STEM (science, technology, engineering, mathematics) disciplines.

The book consists of eight chapters. The first chapter titled *Triangulation in Mathematics and Social Sciences* begins by highlighting ancient techniques of indirect measurement of distances and heights using the idea of triangulation with genesis in work by Thales of Miletus (6th century B.C.). The discussion of the techniques based on the construction of triangles includes finding the height of a pyramid in Egypt, the distance to a boat in the sea from the shore, and other relevant cases of using triangulation as a method of indirect measurement. Basic trigonometric formulas of right triangles are used. The derivation of these and like formulas is discussed in Chapter 2. Several educational documents from North America and elsewhere in the world have been reviewed to show that computational triangulation expands the ideas specific to sociology as using more than one research method, to mathematics with its diverse applications to real life, including education. The expansion from sociology to mathematics education emphasizes more than one way of obtaining an answer to a problem. The chapter briefly mentions how using alternative problem-solving techniques leads to different forms of symbolic answers, especially in trigonometry. A few mathematical examples of using instrumental method in psychology (Vygotsky, 1930) as triangulation have been provided. These examples include using properties of multiplication, the discussion of a conceptually complex idea of the Russian Peasant Multiplication, and methodology of mathematical induction proof with possibilities of computationally supported symbolism of inductive transfer. Finally, the chapter demonstrates the use of triangulation in the study of emotional systems (Bowen, 1994) and shows how geometric ideas integrated with this study can be interpreted in terms of enumerative combinatorics and number theory the models of which allow for the use of computational triangulation. Among digital tools used throughout the chapter are Wolfram Alpha (https://www.wolframalpha.com/), Maple (Char et al., 1991), GeoGebra (Hohenwarter, 2002), and an electronic spreadsheet, as well as the Online Encyclopedia of Integer Sequences (OEIS®, https://oeis.org/).

The second chapter titled *Trigonometry as a Mathematical Basis for Triangulation* discusses how the method of triangulation in mathematics stems from trigonometry and how practical applications of this method are based on different trigonometric formulas. In turn, the derivation of the formulas can display a variety of trigonometric techniques grounded in proving trigonometric identities to enable triangulation in trigonometry itself. The chapter begins with a citation from Bertrand Russell (a brief version of which was included in Chapter 1) to indicate that without knowledge of trigonometry only the case when a ship in the sea was seen by two observers under the angle of 45° can be resolved. It is shown how using basic ideas of trigonometry allows one to find the distance from the land to a ship in the sea in the general case of the location of two observers about the place from where the distance is measured. Other approaches to finding the distance include using the laws of Sines and Cosines. The use of these laws of trigonometry requires proving a sophisticated trigonometric identity by Wolfram Alpha and Maple as a way of computationally triangulating proof in trigonometry. In deriving Ptolemy's identities known from the 2nd century, the chapter shows the use of geometry and, independently, the use of the relationship between complex numbers and trigonometry found by Euler in the mid 18th century. The chapter also illustrates triangulation as using different techniques in solving trigonometric equations and inequalities. Among digital tools used throughout the chapter are Wolfram Alpha, Maple, and the Graphing Calculator (Avitzur, 2011) as a tool of constructing graphs of trigonometric functions, equations, and inequalities.

The third chapter titled *Computational Triangulation, Advanced Conceptual Understanding and TITE Framework* begins with a note that more than one way of providing an answer to a mathematical problem brings about more rigor to problem solving that, in the digital era, may be supported by computational experiments. At the same time, using more than one computer application within the same computational context requires knowledge of specifics of coding used by a particular tool. Towards this end, the discussion includes differences that exist among Wolfram Alpha, Maple, the Graphing Calculator, and a spreadsheet when doing same thing, such as computing or graphing a simple expression. It is suggested that distinguishing between basic conceptual understanding and advanced conceptual understanding in mathematics is the core of theoretical triangulation (Denzin, 1970), something that is considered as the most challenging element of triangulation in sociology. To illustrate, the cases of solving trigonometric and irrational equations are presented to demonstrate that the lack of advanced conceptual understanding may result in the so-called Einstellung effect (Luchins, 1942) or doubled experience (Vygotsky, 1999) in the resulting incomplete solution. Also, the chapter suggests that advanced conceptual understanding of the behavior of functions is required for the accuracy in plotting graphs using software and it shows that a graphing tool might provide inadequate (or simply incorrect) representation of graphs of the functions, even when plotting parabolas symmetrical about a coordinate axis. The technology-immune/technology-enabled (TITE) pedagogical framework is introduced and different mathematical examples of the TITE problem-solving methodology are presented. Among the examples are divisibility property of a fifth-degree polynomial, duality of generalization from limited data when using digital tools capable of symbolic representations of numeric sequences and finding the side lengths of a triangle with the given perimeter. As an illustration of the TITE framework, the last section of the chapter shows how

Binet's formula for Fibonacci numbers to be used in other chapters of the book can be developed.

The fourth chapter titled *Computational Triangulation Using Primary School Context* deals with different scenarios of putting cookies on plates under simple recursive rules. The first scenario follows the rule when the number of cookies beginning from the third plate is the sum of the number of cookies on the previous two plates. In the second scenario, each plate beginning from the third one has twice as many cookies as the first plate of the preceding pair of plates plus the number of cookies on the second plate of the pair. In the third scenario, each plate beginning from the third one has twice as many cookies as the second plate of the preceding pair of plates plus the number of cookies on the first plate of the pair. Each scenario can be described through algebraic sequences the terms of which are linear combinations of the first two terms (the number of cookies on the first two plates) with coefficients defined through linear difference equations of the second order. The transition from the recursive forms to Binet-type closed formulas describing the coefficients is carried out by Maple and Wolfram Alpha. When coefficients in the linear combinations are represented by such (Binet-type) formulas, those combinations of the number of cookies on the first two plates represent the number of cookies on the n-th plate, $n > 2$. Mathematically speaking, the quantities of cookies on plates associated with the first scenario are Fibonacci-like numbers, with the second scenario – Jacobsthal numbers, and with the third scenario – Pell numbers. The Graphing Calculator is used to computationally triangulate the results (that is, in the spirit of McFee (1992), to triangulate between methods) by graphing two straight lines that theoretically are supposed to be identical by describing the number of cookies on the same plate. The first line passes through two points the coordinates of which are represented by the number of cookies on the first two plates, given the number of cookies on the n-th plate. The second line is described by an equation having Binet-type formulas in the left-hand side and the number of cookies on the n-th plate in the right-hand side. Triangulation between methods is also supported by Wolfram Alpha. Several numeric patterns observed within the three scenarios are explained through conceptual remarks. A graphing method is used to carry out triangulation between data sources (Denzin, 1970).

The fifth chapter titled *Computational Triangulation with Cookies on Plates Extended to Fractions* extends explorations of Chapter 4 to the case when fractional parts of cookies can be put on the first two plates to allow for the development of sequences of rational numbers. Consequently, the three scenarios of placing cookies on plates considered in Chapter 4 are discussed under this extension to allow for the modification of recursive equations having unit fractions as the initial values. In turn, mixed fractions as well as integers can represent the quantities of cookies on other plates. Several approaches to modeling the number of cookies on plates within the first scenario are implemented and computational triangulation is used to demonstrate that symbolic results provided by different digital tools are identical. A new focus of explorations includes the study of the transient time of moving from plate to plate having the whole number of cookies. Whereas explorations using different tools confirm each other results, the results appear not demonstrating any patterns. The chapter shows how four digital tools – Maple, Wolfram Alpha, a spreadsheet, and the Graphing Calculator – can be used to model Binet-like formula as the solution of a linear recursive equation with the initial values $1/k$ and 1 (as the quantities of cookies on the first and the second plates, respectively) to generate the sequences of rational numbers for different values of k. For example, all four tools confirm that in the context of the first scenario, when $k = 6$, the 14th plate, having 257 cookies, is the first one after the second plate with the whole number of cookies. One can see that the transient time between the second and the 14th plates is measured by 11 plates, and this number of plates apparently stays the same, as one moves further up along the plates to reach a plate with a whole number of cookies. In the context of the first scenario, the transient time is connected to divisibility of Fibonacci numbers, serving as coefficients in the linear combinations of the number of cookies on the first two plates, by k. In particular, it is shown that among any three consecutive Fibonacci numbers only one number is even. Wolfram Alpha is used to generate the remainders of the first 50 Fibonacci numbers when divided by k, $2 \leq k \leq 7$. Same explorations are carried out in the context of the second and the third scenarios.

The sixth chapter titled *Computational Triangulation Using Secondary School Context* continues geometric explorations of Chapter 1 regarding Bowen's emotional system. In this chapter, the system is expanded from triangles to polygons. It begins with the question: How many convex quadrilaterals can one construct by connecting four, five, six, and so on individuals associated with an emotional system? The notion of computational triangulation is applied towards the development of a sequence representing the number of quadrilaterals formed by an m-person emotional system, $m \geq 4$. The summation formulas for consecutive natural, triangular, tetrahedral, and pentatope numbers along with the rising factorial function (the Pochhammer symbol) are considered to demonstrate simple yet conceptually profound connections between those numbers. A problem about triangle and its perimeter with the side lengths being consecutive triangular numbers is formulated and solved using the triangle inequality expressed through the Pochhammer symbols. Alternatively, it is shown how the problem can be solved visually by the geometrization of arithmetic. A similar problem is considered when three consecutive tetrahedral numbers serve as the side lengths of a triangle. Through these explorations, different formulas connecting triangular and tetrahedral numbers have been developed. The chapter concludes with using four digital tools – a spreadsheet, Wolfram Alpha, Maple, and the Graphing Calculator – as a way of computationally triangulating different symbolic results developed through the perimeter of triangle problems.

The seventh chapter titled *Computational Triangulation in Elementary Number Theory* explores subsequences of polygonal numbers of different sides derived through step-by-step elimination of the terms of the original sequence. Eliminations follow special rules similarly to how the classic sieve of Eratosthenes was developed through eliminating the multiples of prime numbers. The chapter begins with highlighting various representations of numbers through the sums of other numbers as one of the big ideas of mathematics. Among these ideas are representation of squares as sums of odd numbers, Pythagorean theorem, Goldbach's conjecture, Tertiary Goldbach's conjecture, and representation of numbers as sums of polygonal numbers. The notion of a sieve of order k of triangular numbers is introduced along with three problem-solving strategies used to mathematise computational inquiries into such sieves. The first strategy is based on the use of Wolfram Alpha to generalize from the first few terms of a triangular number sieve to obtain its algebraic representation. A general formula of the triangular number sieve of order k is developed in the form of a second-degree polynomial (in terms of the rank of the original triangular number sequence) the coefficients of which are the powers of two of the sieve's order. The approach is based on the TITE framework which is interpreted in terms of the notion of the instrumental act discussed in Chapter 1. The second strategy in the development of formulas for triangular number sieves is based on the study of the ranks of triangular numbers involved in a sieve. These ranks include the powers of two of the sieve's order. Then, computational triangulation is used to demonstrate that symbolically different formulas for triangular number sieves of order k developed independently by Wolfram Alpha and Maple are identical. The third problem-solving strategy is to use a spreadsheet to verify the accuracy of symbolic representations provided by Maple and Wolfram Alpha and then generalize the results obtained through the second method to include a formula of the sieve of polygonal numbers of an arbitrary side. In the last section, using the TITE framework, polygonal number sieves of the Cullen type are considered by interpreting the results of computations carried out in the previous three sections in mathematical terms.

The eighth chapter titled *TITE Explorations of Two Discrete Mathematics Topics with Historical Flavor* starts with a classic problem of finding the probability of irreducibility of a fraction with randomly selected numerator and denominator from the set of natural numbers. This problem may be safely regarded as one of the most famous problems situated as the confluence of the theory of probability and the theory of numbers. A technology-immune (TI) part of explorations begins with the context of spinners divided into eight (even) and nine (odd) equal sections numerated by the first eight and nine natural numbers, respectively, and finding the probability of landing on an even number for each spinner. As a technology-enabled (TE) part of the activities, a spreadsheet is used to calculate relative frequency of selecting multiples of a prime number p (the number 5 is used as an example) among 10,000 randomly generated integers in the range $[2, 10^4]$ to enable a heuristic conclusion that, in general, the probability (understood as relative frequency) of selecting multiples of p among the set of natural numbers is equal to $1/p$. The discussion continues in a TI way by finding the probability of selecting two natural numbers not having a prime number p as a common factor leading to Euler's famous identity connecting an infinite product (involving the reciprocals of the squares of primes) to an infinite sum (involving the reciprocals of the squares of natural numbers) through which the probability sought has been found. A spreadsheet is used to confirm the theoretical result experimentally through five million trials. Also, by interpreting irreducible fractions as lattice points with relatively prime coordinates and expressing the number of such points within a large square in the first quadrant through the Euler's phi function, this TI result is confirmed through a TE computational triangulation. As the result, the famous probability $\frac{6}{\pi^2} \cong 0.608$ obtained through heuristic reasoning is confirmed experimentally using the power of digital tools. The second topic discussed in the chapter makes it possible to see how the asymmetry of entries of the classic Pascal's triangle can be used in a TI fashion to develop Fibonacci-like polynomials, the coefficients of which add up to Fibonacci numbers, two polynomials of each degree. Exploring those polynomials by digital tools leads to the concept of generalized Golden Ratios in the form of strings of real numbers of different lengths defined by equations with continued fractions. Connection of continued fractions and Fibonacci-like polynomials is demonstrated. It is shown that the roots of the polynomials are all real numbers located within the interval (-4, 0). Computations and graphing are used to describe symmetrical and asymmetrical location of the roots within that interval. Among digital tools used in this chapter, in addition to a spreadsheet, are Wolfram Alpha, Maple, the Graphing Calculator and GeoGebra.

Chapter 1

Triangulation in Mathematics and Social Sciences

1.1. Introduction

Triangle is the basic geometric figure the properties of which have been used in real-life applications of mathematics since ancient times. Its shape appears in the Egyptian Papyrus Roll (1650 B.C.), known as the Rhind mathematical papyrus (Chace et al., 1927), resembling an isosceles triangle the height of which is visually indistinguishable from a lateral side. This type of triangle allowed ancient Egyptians to calculate its area by replacing isosceles triangle by rectangle the area of which (visually) is twice the area sought (Boyer & Merzabach, 1989). In turn, the genesis of emphasizing isosceles triangle in the papyrus goes back to the architecture of Neolithic age (10,200 B.C. – 2000 B.C.) when houses had roofs in the form of an isosceles triangle with lateral sides almost touching the ground (Kuijt, 2002). Thousand years after the Egyptians, a Greek mathematician Thales from Miletus (6-5 centuries B.C.), a pre-Socratic Greek philosopher and one of the founders of ancient Greece, used knowledge of similarity of right triangles to measure indirectly large heights (e.g., pyramids in Egypt) and long distances (e.g., in seas or land with impassable obstruction between two points).

The idea of triangulation resides in using similarity of triangles and its extension to trigonometry in indirect measurement of the heights and distances. Figure 1.1 shows how Thales, according to historical records, found the height (H) of a pyramid in Egypt by knowing the length of its shadow (S) and his own height (h_1) and shadow (s_1). According to Russell (1945), Thales "seems to have discovered … how to estimate the height of a pyramid from the length of its shadow" (p. 25). The similarity of triangles ABC and ADE yields $\frac{H}{S} = \frac{h_1}{s_1}$ whence $H = S\frac{h_1}{s_1}$. The last relation yields $H = S$ when $h_1 = s_1$, i.e., when someone's height and shadow are equal. Similar problems of practical mathematics using the ideas of triangulation like finding the width of a river or the distance between two points blocked by impassable barrier were also solved in the 6th century B.C. by scholars of Islamic civilization (Hill, 2000).

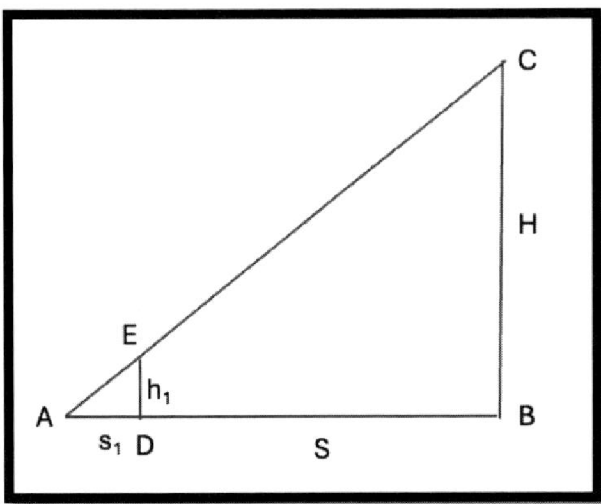

Figure 1.1. Finding the height of a pyramid.

1.2. From Mathematics to Sociology in the Use of Triangulation

A brief historical introduction makes it clear that the word triangulation, whatever its meaning, has mathematical origin. One such meaning, promoted in this book, deals with the modern-day use of digital tools with distinct computational features. As was mentioned in Preface, computational triangulation is a new term that grew out of the author's experience using computers with future K-12 teachers of mathematics and interest in learning how social sciences research was enriched by the triangulation approach to sociology. To this end, the link between mathematics education and sociology was developed (Abramovich, 2022). The use of sophisticated and diverse software tools by teacher candidates encourages the development of deep mathematical knowledge required nowadays for teaching mathematics across the grades. Furthermore, Van Bendegem's (1998) perspective on a mathematical experiment as the integration of digital computation and physical manipulation makes it possible, by using concrete materials to enhance the visual component of mathematics education, including the mathematical preparation of schoolteachers. Indeed, as mentioned by the Conference Board of the Mathematical Sciences (2012), an umbrella organization of twenty professional societies in the mathematical sciences in the United States, concerned, in particular, with the mathematical preparation of K-12 mathematics teachers, "Software, manipulatives, and many other tools exist to support teaching and learning … and to help students use them strategically in doing mathematics, teachers need to understand the mathematical aspects of these tools and their uses" (p. 2). Likewise, mathematics educators in Ontario province of Canada remind us that "computational thinking, coding, design thinking, scientific inquiry skills … are in high demand in today's globally connected world, with its unprecedented advancements in technology" (Ontario Ministry of Education, 2020, p. 30). At the same time, it is worth noting that some eight thousand miles from North America, mathematics educators in South Africa argue "that mathematics teachers, and not ICT tools, are the key to quality education" (Department of Basic Education, 2018, p. 78). This is why the author considers the concept of computational triangulation, along with everything that, as educators in England mentioned, emphasizes "more than one way of doing things" (Advisory Committee on Mathematics Education, 2007, p. 18), being an important focus of the modern-day mathematics teacher education courses. In Korea, an emphasis on doing mathematical and technological things differently is connected to the development of upright character through "activities that lead students to respect others' way of solving problems and others' ideas in order to build a character that cares others" (Park & Choi-Koh, 2013, p. 99). In Australia, mathematics educators encourage schoolteachers' appreciation that "digital technologies … allow new approaches to explaining and presenting mathematics … support numerical, statistical, graphical, symbolic, geometric and text functionalities … allow greater attention to meaning, transfer, connections and applications" (National Curriculum Board, 2008, p. 9).

Seeing triangulation as the use of multiple methods in sociological research, like multiple digital tools in computation, Denzin (1970) referred to this concept as "a plan of action that will raise sociologists above the personalistic biases that stem from single methodologies" (p. 300). Whereas triangulation in the social science research is a relatively new concept aimed at improving the validity of findings based on alternative epistemologies in the study of the same object (Campbell & Fiske, 1959; Webb et al. 1996), the idea of validation of mathematical propositions, going back to antiquity (Kitcher, 1983), and/or simplification of their proofs using alternative reasoning techniques, including computational mathematics, is a commonplace of pure and applied mathematics research (e.g., Metropolis & Ulam, 1949; Lauder & Parkin, 1966; Elkies, 1988; Grinshpan, 1999; Leonov & Kuznetsov, 2013; Bailey & Borwein, 2018). Löwe & van Kerkhove (2019) use the term triangulation to discuss from a philosophical perspective the issue of confidence of research mathematicians in the correctness of proof of a mathematical proposition. This issue was discussed by Harrison (2008) who argued that "the correctness of mainstream mathematical proof is almost never established by formal means, but rather by informal discussion between mathematicians and peer review of papers" (pp. 1398, 1399). In mathematics education, formal proof is typically understood in terms of Pólya (1954) who writes about "formal or demonstrative logic" (p. v) and "formal proof" (p. 51). Half a century after Pólya, Harrison's (2008) use of the words formal proof means "a proof written in a precise artificial language that admits only a fixed repertoire of stylized steps" (p. 1395). This makes the issue of triangulation as a way of endowing rigor in proving a mathematical proposition rather complicated and the notion of computational triangulation discussed in this book is something between Pólya's and Harrison's perspectives on mathematical rigor in formal demonstration (see also Chapter 7, Section 7.1, about Archimedes perspective on mathematical proof).

Sharma (2013) considered the concept of triangulation by pointing at the limitations of the interview-based qualitative methods in mathematics education research. Mok and Clarke (2015) used the idea of triangulation to describe and accommodate in mathematical education research the cultural complexity of classroom pedagogy within Asian and non-Asian mathematics teaching contexts. Bachman et al. (2020) used the term triangulation by referring to the collection of a variety of qualitative data regarding parental influence on the development of mathematical skills in the early childhood. Consistent with a view that "the term 'triangulation' is used in many different ways" (McFee, 1992, p. 215), one usage of the term (introduced in this book) may be associated with mathematics teacher education to be understood as using multiple ways of solving a problem (Abramovich, 2021), often enabling a problem solver, from a pure pragmatic perspective, to "check the result" (Pólya, 1957, p. 59) by arriving to the same answer from different directions; that is, using different methods of solving a problem. This perspective develops through practicing productive thinking (Wertheimer, 1959) and recognizing the value of insight in problem solving (Dunker, 1945) which is akin to the act of sagacity defined by Aristotle as "hitting by guess upon the essential connection in an inappreciable time" (cited in Pólya, 1973, p. 58). Sometimes, alternative problem-solving methods produce different forms of answers (e.g., in trigonometric equations and inequalities or in plane and solid geometry where trigonometry is used) leading to "triangulation *between* methods … [and] triangulation *within* a method" (McFee, 1992, p. 217; italics in the original). By posing and solving new problems that deal with comparison of symbolically different answers and proving their numerical equivalence, one applies computational triangulation *between* problem-solving methods. In order to deal with alternative problem-solving strategies, a student must understand mathematics behind the concepts involved, to be able "to justify in a way appropriate to the student's mathematical maturity, *why* a particular mathematical statement is true or where a mathematical rule comes from" (Common Core State Standards, 2010, p. 4, italics in the original), and to be given opportunities "to experience the living practice of mathematics, to work through challenges, and to find success and beauty in problem solving" (Ontario Ministry of Education, 2020, p. 64). In South America, Chilean mathematics educators recommend presenting a mathematical concept through multiple representations because it "helps students achieve a deeper understanding of the concept and provides more tools for working with it" (Felmer et al. 2014, p. 35). In Asia, a similar recommendation is provided by Korean mathematics educators who argue that already in grade school "using a variety of reasoning abilities to understand ideas … results in more flexible thinking and establishes deeper understanding through connecting various ideas in multiple ways" (Cho & Kim, 2013, p. 131). Likewise, a similar perspective can be found in Japanese mathematics education materials suggesting that through the interplay between mathematics and digital computations "students will engage actively in mathematical activities … and deepen their understanding of relations between mathematics and scientific technology" (Isoda, 2010, p. 13). Towards this end, future teachers of mathematics must be educated to possess deep conceptual understanding of the subject matter, recognize the value of technology as a representational tool, and be supported in developing "the habits of mind of a mathematical thinker and problem-solver, such as reasoning and explaining, modeling, seeing structure, and generalizing" (Conference Board of the Mathematical Sciences, 2012, p. 19). In the age of technology, when mathematical problem solving involves more than one digital tool, another type of triangulation – computational triangulation – can be used in mathematics education, in general, and in its teacher education component, in particular, enabling students "to use technological tools to explore and deepen their understanding of concepts" (Common Core State Standards, 2010, p. 7). To achieve such outcome, students need teachers, who through their own studies of mathematics "develop rigorous habits of mind through mathematical reasoning and proof" (Ministry of Education Singapore, 2012, p. 8).

1.3. The Instrumental Method in Psychology as a Triangulation

The idea of triangulation can be found in the writings of Lev Vygotsky on psychology about the instrumental method used in a variety of situations, problem solving included. Simply put, an instrumental method incorporates an artificial psychological tool placed between a problem to be solved and a straightforward method of solution to modify the relation between the two. According to Vygotsky (1930),

> "The inclusion of a tool in the behavioral process, first, sets to work a number of new functions connected with the use and control of the given tool; second, abolishes and makes unnecessary a number of natural processes, whose work is [now] done by the tool; third, modifies the course and the various aspects (intensity, duration, order, etc.) of all mental processes included in the instrumental act, replacing some functions with others, i.e., it recreates, reconstructs the whole structure of behavior just like a technical tool recreates the entire system of labor operations. Mental processes, taken as a whole, form a complex structural and functional unity. They are directed toward the solution of a problem posed by the object, and the tool dictates their coordination and course. They form a new whole – the instrumental act."

To illustrate the notion of the instrumental act in mathematics through the lens of triangulation, three examples will be provided: commutative and distributive properties of multiplication, the Russian Peasant Multiplication, and mathematical induction proof.

1.3.1. Using Properties of Multiplication as an Instrumental Act

Consider the use of commutative property of multiplication as a psychological tool which can enhance computing the product 10×7 by a child without recourse to a calculator. The product of two factors means that the first factor shows the number of times the second factor is repeated through addition. That is, the path from the product 10×7 to the final answer, 70, is through the straightforward calculation of the sum $\underbrace{7 + 7 + \cdots + 7}_{10\ times}$. However, there is no easy strategy of counting by sevens, and the commutative property of multiplication is inserted as an artificial tool between the product 10×7 and the answer 70 obtained through adding ten sevens. In that way, if the stimulus A is the product 10×7 and the stimulus B is the answer 70 $(= \underbrace{10 + 10 + \cdots + 10}_{7\ times})$, according to Vygotsky (1930), the equality $10 \times 7 = 7 \times 10$ is a psychological tool C inserted between A and B forming the triangle ACB shown in Figure 1.2. In that way, instead of counting by sevens in the transition from A to B, we first create the path from A to C followed by the path from C to B. That is, $10 \times 7 = 7 \times 10 = 10 + 10 + 10 + 10 + 10 + 10 + 10 = 70$.

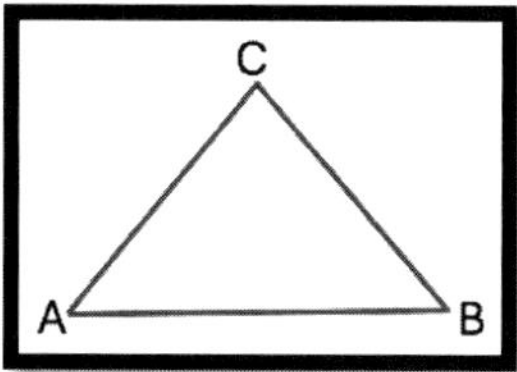

Figure 1.2. Instrumental act as a triangulation.

Another example is finding the square of the number 51 by using a diagram which reduces squaring to finding areas of rectangles representing the distributive property of multiplication over addition. Figure 1.3 shows how squaring can be presented as an instrumental act by placing this property (enhanced by its visual representation) between the square 51^2 and the answer 2601.

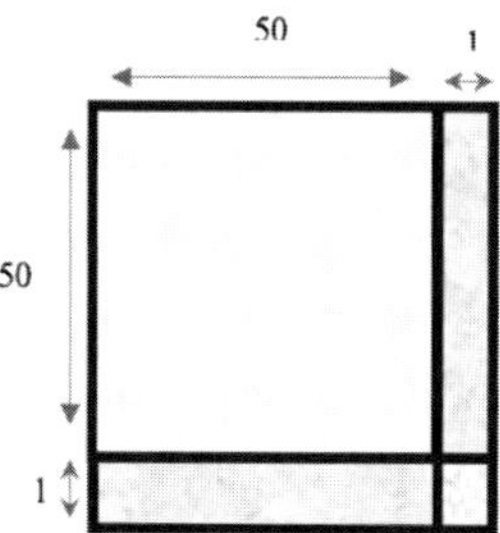

Figure 1.3. Decomposing 51 into 50 + 1 as an instrumental act.

As shown in Figure 1.3, the decomposition 51 = 50 +1 simplifies computation as follows:

$51^2 = (50 + 1)^2 = 50 \times 50 + 50 \times 1 + 1 \times 50 + 1 \times 1 = 2500 + 100 + 1 = 2601.$

Once again, the process of using the distributive property (an abstract concept) can be described as a triangle when the squaring of the number 51 (vertex A in Figure 1.2) leads to the result (vertex B), 2601, through vertex C which is an artificial tool used to facilitate calculation (as shown in Figure 1.3). Alternatively, one can use a calculator (and place it in vertex C) as the replacement of the diagram of Figure 1.3 in finding the square of the number 51.

1.3.2. The Russian Peasant Multiplication as an Instrumental Act

The square of the number 51 can be found by using the Russian Peasant Multiplication (Gimmestad, 1991; Cevizci, 2018; Mironov, 2019). This algorithm of finding the product of two factors is based on the idea that every natural number can be represented as a sum of distinct powers of two. For example, $51 = 32 + 16 + 2 + 1 = 2^5 + 2^4 + 2^1 + 2^0$ (and not $2^4 + 2^4 + 2^4 + 2^1 + 2^0$ as the powers of two here are not distinct), $31 = 16 + 8 + 4 + 2 + 1 = 2^4 + 2^3 + 2^2 + 2^1 + 2^0$, and $30 = 16 + 8 + 4 + 2 = 2^4 + 2^3 + 2^2 + 2^1$. It is the presence or the absence of the term 2^0 that points at the parity of a number, odd (like 51 and 31) or even (like 30). Furthermore, one can see that in the representation of 31 all powers of two smaller than 31 are present and in the representation of 51 the powers 2^3 and 2^2 are missing. In order to find the product 51×51 using the Russian Peasant Multiplication; in other words, to find the product $(2^5 + 2^4 + 2^1 + 2^0) \times 51$, one creates a three column table (Figure 1.4) with the numbers $2^0, 51$, and 51 at the top of each column, respectively. Then, in the far-left column one continuously increases powers of two from 0 to 5; in the middle column continuously divides numbers by 2 and record the integer part of the division; in the far-right column continuously multiply numbers by 2, and finally add all numbers in the far-right column that correspond to the powers of two present in 51; that is, not including those in the rows for 2^2 and 2^3. One can note that only for those two cases the numbers in the middle column are even and many online sources of information associated with the Russian Peasant Multiplication instruct (procedurally, without explanation) crossing out the corresponding numbers in the far-right column in order to be excluded from the addition – see below Remark 1.1 for a conceptual explanation of this procedural rule. Once again, we have $51 \times 51 = 51 \times (2^0 + 2^1 + 2^4 + 2^5) = 51 + 102 + 816 + 1632 = 2601$.

2^0	51	51
2^1	$25 = INT(51 \div 2)$	$102 = 51 \times 2$
2^2	$12 = INT(25 \div 2)$	$204 = 102 \times 2$
2^3	$6 = INT(12 \div 2)$	$408 = 204 \times 2$
2^4	$3 = INT(6 \div 2)$	$816 = 408 \times 2$
2^5	$1 = INT(3 \div 2)$	$1632 = 816 \times 2$

Figure 1.4. Using the Russian Peasant Multiplication to find 51×51.

Figure 1.5 shows finding the product 31×5 using the Russian Peasant Multiplication. One can see that continuously dividing $31 = 2^4 + 2^3 + 2^2 + 2^1 + 2^0$ by 2 and taking the integer part of the division yield only odd numbers and therefore all numbers from the far-right column are included in the sum. Indeed,

$$INT\left(\frac{31}{2}\right) = 2^3 + 2^2 + 2^1 + 1;\ INT\left(\frac{31}{2^2}\right) = 2^2 + 2^1 + 1; INT\left(\frac{31}{2^3}\right) = 2^1 + 1;\ INT\left(\frac{31}{2^4}\right) = 1.$$

At the same time, in the case of $51 = 2^5 + 2^4 + 2^1 + 2^0$, continuous division by two and the use of the INT function yield

$$INT\left(\frac{51}{2}\right) = 2^4 + 2^3 + 2^1 + 1;\ INT\left(\frac{51}{2^2}\right) = 2^3 + 2^2 + 1; INT\left(\frac{51}{2^3}\right) = 2^2 + 2;\ INT\left(\frac{51}{2^4}\right) = 2;$$

$INT\left(\frac{51}{2^5}\right) = 1$, that is, only divisions by 2, 2^2, and 2^5 yield odd values of the greatest integer function (see Figure 1.5). Note that the number $5 = 2^2 + 2^0$ could be put in the middle column and, in that case, the product $31 \times 5 = 31 + 124 = 155$. So, it is immaterial which of the two factors to place in which column.

20	31	5
21	$15 = INT(31 \div 2)$	$10 = 5 \times 2$
22	$7 = INT(15 \div 2)$	$20 = 10 \times 2$
23	$3 = INT(7 \div 2)$	$40 = 20 \times 2$
24	$1 = INT(3 \div 2)$	$80 = 40 \times 2$

Figure 1.5. Using the Russian Peasant Multiplication to find 31×5.

	A	B	C	D	E	F	G	H	I	J
1										
2		8	9	10	11	12	13	14	15	16
3	1	0	1	0	1	0	1	0	1	0
4	2	0	0	1	1	0	0	1	1	0
5	4	0	0	0	0	1	1	1	1	0
6	8	1	1	1	1	1	1	1	1	0
7	16	0	0	0	0	0	0	0	0	1

Figure 1.6. Using a spreadsheet to model Russian Peasant Multiplication between 8 and 16.

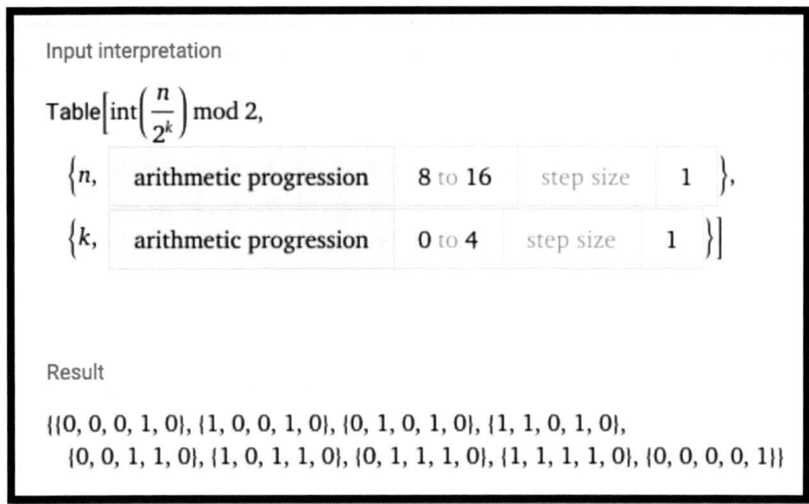

Figure 1.7. Wolfram Alpha modeling Russian Peasant Multiplication in the range $8 < n < 16$.

One can use a spreadsheet (Figure 1.6) and Wolfram Alpha (Figure 1.7) to model the Russian Peasant Multiplication when one of the factors is relatively small. For example, consider the product $10 \times N$. In column D (Figure 1.6) we have two ones, in cells D4 and D6. The corresponding numbers in cells A4 and A6 are 2 and 8. This leads to the sum 2N + 8N = 10N – exactly the product $10 \times N$. The same result can be obtained by looking at the third string {0, 1, 0, 1, 0} generated by Wolfram Alpha (Figure 1.7) and representing $10 = 2 + 8$. The zeroes in this string are associated with the powers of two: 2^0, 2^2, and 2^4. So, once again, the product $10 \times N$ can be found as the sum of $2^1N + 2^3N = 10N$

Remark 1.1. In order to explain why the products corresponding to even integer values of continuous divisions of the factor in the middle column are not included in the final sum as shown in Figure 1.4, several things must be stated. First, as was mentioned above, any natural number can be represented as a sum of distinct powers of two. Just as powers of ten are the place values in base ten, (e.g., $51 = 5 \times 10^1 + 1 \times 10^0$ or $501 = 5 \times 10^2 + 0 \times 10^1 + 1 \times 10^0$), powers of two are the place values in the binary system which has two digits only, 0 and 1 (e.g., $51 = 1 \times 2^5 + 1 \times 2^4 + 0 \times 2^3 + 0 \times 2^2 + 1 \times 2^1 + 1 \times 2^0$, thereby, not having same powers of two). Gottfried Wilhelm Leibniz (1646-1716), the great German mathematician and philosopher, was one of the early modern era proponents of the binary system. However, the genesis of the binary system goes back to ancient (the 3rd century B.C.) China to emphasize a dichotomy of the "yes-no" or the "on-off" dualities. Second, as was already mentioned, those representations in which 2^0 (= 1) is present are odd numbers, otherwise the number is even. Third, for any natural number $N \geq 2$, there exists natural number $n = INT(log_2 N)$ so that 2^n is the largest power of two included in the representation of N as a sum of distinct powers of two. For example, $INT(\log_2 51) = 5$. Fourth, when in the representation of N as a sum of distinct powers of two the power 2^k, $k < n$, is absent, then the Russian Peasant Multiplication of the product $N \times M$ would not include the term $2^k \times M$ in the far-right column. By the same token, the subsequent division of $N = 1 \times 2^n + 1 \times 2^{n-1} + \cdots + 1 \times 2^{k+1} + 0 \times 2^k + 1 \times 2^{k-1} + \cdots$ by 2 on step k would yield (as the integer part) $2^{n-k} + 2^{n-k-1} + \cdots + 2^1$, thus not including 2^0, the only power of two responsible for a number being odd. For example, in the representation of 51 the powers 2^2 and 2^3 are absent. This means that the products $2^2 \times M$ and $2^3 \times M$ will not be included in the sum of products appearing in the far-right column. For example, when $N = 51$ and $M = 100$ the resulting sum for the product 51×100 would be equal to $(2^0 + 2^1 + 2^4 + 2^5) \times 100 = 2^0 \times 100 + 2^1 \times 100 + 2^4 \times 100 + 2^5 \times 100 = 100 + 200 + 1{,}600 + 3{,}200 = 5{,}100$.

1.3.3. Mathematical Induction Proof as an Instrumental Act

The third example of the instrumental act that "recreates, reconstructs the whole structure of [mathematical] behavior" (Vygotsky, 1930) in dealing with infinite processes is mathematical induction proof. The instrument of proof is the transition from n to $n + 1$ in proving a statement $P(n)$ depending on n. This transition allows one to prove that if the statement $P(n)$ is true, then the statement $P(n + 1)$ is also true and there is no need to do proof for every n within the infinite set of natural numbers. To illustrate, consider proving the statement that the sum of the first n odd numbers is equal to n^2, that is (see also the beginning of Chapter 7),

$$1 + 3 + 5 + \cdots + 2n - 1 = n^2. \quad (1.1)$$

Assuming that equality (1.1) is true, one increases its left-hand side by the next odd number to have the sum $(1 + 3 + 5 + \cdots + 2n - 1) + (2n + 1)$, which, due to (1.1), is equal to $n^2 + 2n + 1$. The last trinomial can be represented as $(n + 1)^2$ using the instrument of Figure 1.3 in which 50 is replaced by n.

Mathematical induction proof can be supported by digital symbolic computations when algebraic transformation needed to demonstrate the transition from n to $n + 1$ is too complicated to do it unlike the case of proving identity (1.1). As an example, consider proving divisibility by 12 of the polynomial $R(n) = n^2(n + 1)^2(2n^2 + 2n - 1)$ for all $n = 1, 2, 3, \ldots$ (see Chapter 3, Section 3.5). The base clause of the induction is true as $R(1) = 12$. Assuming that $R(n)$ is divisible by 12, one can use Maple (Figure 1.8) to show that the difference $R(n + 1) - R(n)$ is divisible by 12 as well. Because Maple, by doing necessary symbolic computations, yields the relation $R(n + 1) - R(n) = 12(n + 1)^5$, the inductive assumption about $R(n)$ and the right-hand side of the last relation being an obvious multiple of 12 imply that $R(n + 1)$ is divisible by 12 as well. In the spirit of computational triangulation, similar symbolic computations can be outsourced to Wolfram Alpha (Figure 1.9). This completes the mathematical induction proof independently supported by two digital tools. In the words of Langtangen and Tveito (2001), "Much of the current focus on algebraically challenging, lengthy, error-prone paper and pencil work can be significantly reduced. The reason for such an evolution is that the computer is simply much better than humans on any theoretically phrased well-defined repetitive operation" (pp. 811-812). Nowadays, this position is fully applied to the mathematical induction proof as an instrumental act.

> $R(n) := n^2 \cdot (n+1)^2 \cdot (2 \cdot n^2 + 2n - 1)$

$$R := n \mapsto n^2 \cdot (n+1)^2 \cdot (2 \cdot n^2 + 2 \cdot n - 1)$$

> $R(n+1)$

$$(n+1)^2 (n+2)^2 (2(n+1)^2 + 2n + 1)$$

> $R(n+1) - R(n)$

$$(n+1)^2 (n+2)^2 (2(n+1)^2 + 2n + 1) - n^2 (n+1)^2 (2n^2 + 2n - 1)$$

> $simplify(\%)$

$$12(n+1)^5$$

Figure 1.8. Maple-supported mathematical induction proof.

Input interpretation

$$P(n+1) - P(n) \text{ where } P(n) = n^2 (n+1)^2 \left(2n^2 + 2n - 1\right)$$

Result

$$(n+1)^2 (1 + (n+1))^2 \left(-1 + 2(n+1) + 2(n+1)^2\right) - n^2 (1+n)^2 \left(-1 + 2n + 2n^2\right) = 12(1+n)^5$$

Figure 1.9. Wolfram Alpha carries out the transition from n to $n + 1$.

1.4. Triangulation in the Study of Emotional Systems

Murray Bowen's theory of family system functioning includes the concept of triangulation as a theoretical framework for understanding how all emotional systems function. According to Bowen (1994), an emotional system of two individuals is unstable and "it forms itself into a three-person or triangle under stress" (p. 478). The role of a triangle formed by three individuals is to relieve stress that may result from the interaction of any two members of the triangle. One such example is a tension between two siblings when their mother, having positive relation with each child can "triangle in" (ibid, p. 478) herself to relieve the stress between her kids. In other cases, when more than three individuals are involved in the emotional situation, the concept of triangulation is understood as a series of interlocking triangles.

According to Bowen (1994), a triangle connecting three individuals (one of whom is considered an outsider), say, mother, child, and father, has two positive sides and one negative side. Such triangle can be described as an isosceles triangle with the base being a negative side. If such triangle is fluid, it tends to form an isosceles shape with the length of the base dependent on the magnitude of the negative relation between two actors. The triangle inequality, stating that the sum of any two sides of a triangle is larger than the third side (see also Chapter 3, conclusion of Section 3.5, and Chapter 6, Section 6.4), suggests that negativity may not be larger than or equal to combined positivity of other two relations. In other words, the extent of negativity may not be too high otherwise the triangle representing an emotional system does not exist. That is, an emotional system with a high magnitude of negativity between two actors cannot be described through a triangle.

Bowen (1994) provided a relationship between the number of people involved in an emotional system and the number of triangles associated with the system: "A three-person system is one triangle, a four-person system is four primary triangles, a five-person system is *nine* primary triangles, etc. This progression multiplies rapidly as systems get larger" (p. 479, italics added). A typo (italicized) in this quote can be used to demonstrate the idea of triangulation in mathematics as more than one way of solving a problem and fill in for the "etc." in the last quote.

One can see (Figure 1.10) that by adding to a three-person system (forming a single triangle) another person D, adds three new triangles: ABD, ACD, and BCD. Figure 1.11 shows that adding to a four-person system ABCD (forming four triangles: ABC, ABD, ACD, and DBC) another person E, adds six new triangles: ABE, AEC, DBE, DEC, CBE, and EBC. This makes the total of ten (not nine) triangles for a five-person system.

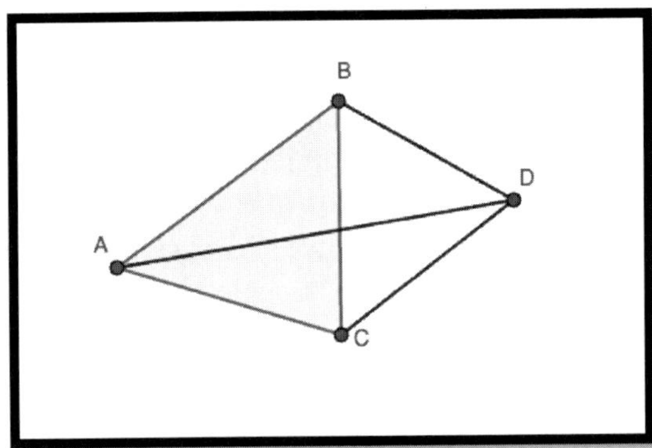

Figure 1.10. Four triangles in a four-person system.

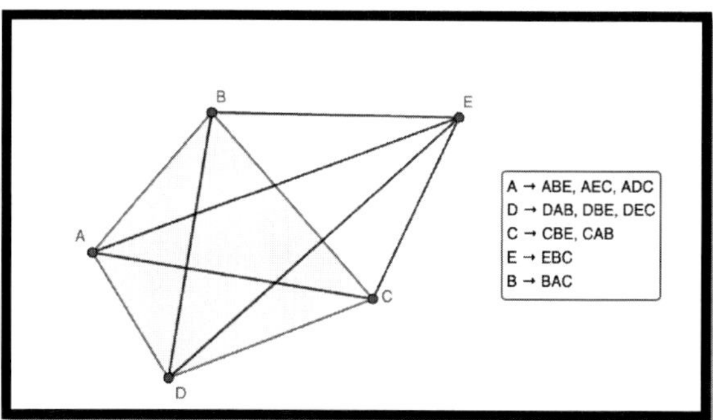

Figure 1.11. Ten triangles in a five-person system.

One can also show the presence of typo in the above quote by making an organized list of triangles out of five vertexes A, B, C, D, and E. The first six triangles include vertexes (persons) A and B: ABC, ABD, ABE; then vertexes (persons) A and C (without B): ACD, ACE; then vertexes (persons) A and D (without B and C): ADE. The remaining four triangles are BCD, BCE, BDE, and CDE. This makes the total number of triangles equal ten.

Another way to show that a five-person system forms ten triangles is to find the number of ways of selecting three different objects (vertexes) out of five different objects (persons). In other words, through this selection process one creates what is called a three-combination of five objects. This selection is described by the formula $C_5^3 = \frac{5!}{3!(5-3)!} = 10$. Once again, one can see that a five-person system forms ten (not nine) triangles. To explain the last formula, note that any three-combination of five objects may be described through a five-letter word with three letters Y pointing at the objects included in a combination and two letters N pointing at the objects not included in that combination. Therefore, the number of permutations of letters in the word YYYNN is equal to the number of three-combination of five objects. One can prove by the method of mathematical induction that the number of permutations of n different letters is $n!$. Indeed, when $n = 1$, one letter can be permuted in one way only and $1! = 1$. Assuming that n letters can be permuted in $n!$ ways, within each such permutation the $(n + 1)$-st letter can be placed in $n + 1$ ways. Therefore, by the Rule of Product stating that if object A can be selected in m ways and if, following this selection, object B can be selected in n ways, then the *ordered* pair (A, B) can be selected in mn ways, the number of permutations of letters in a word with $n + 1$ letters is equal to the product $n! \times (n+1) = (n+1)!$. In particular, when $n = 5$ we have 5! permutations of different letters in a five-letter word. Let P be the number of permutations in the word YYYNN. Noting that in each such permutation letters Y can be permuted in 3! ways and letters N can be permuted in 2! ways (where 2! = (5 – 3)!), by the Rule of Product the number of permutations of letters in the word YYYNN when all letters may be permuted is $P \times 3! \times (5-3)!$. Therefore, $P \times 3! \times (5-3)! = 5!$ Whence $P = \frac{5!}{3! \times (5-3)!} = 10$.

In much the same way, by permuting letters in an m-letter word $\text{YYY}\underbrace{\text{NN} \ldots \text{N}}_{m-3}$, the general formula $C_m^3 = \frac{m!}{3!(m-3)!}$ can be developed and used within a spreadsheet to generate the number of triangles (each defined by three vertexes) formed by an m-person system as m increases. Such a spreadsheet is shown in Figure 1.12. In the spirit of computational triangulation, one can use Wolfram Alpha (Figure 1.13) to generate same numbers, thereby, filling for the "etc." in the above quote from Bowen (1994).

	A	B	C	D	E	F	G	H	I	J	K	L	M
1	persons	3	4	5	6	7	8	9	10	11	12	13	14
2	triangles	1	4	10	20	35	56	84	120	165	220	286	364
3													

Figure 1.12. Using a spreadsheet in filling for the "etc." in the quote from Bowen (1994).

Alternatively, the last formula for C_m^3 can be developed using the Rule of Sum: if object A can be selected in k ways and, independently from this selection, object B can be selected in n ways, then there are $k + n$ ways to select either object A or object B. To this end, one can put aside the first object and select three objects out of the remaining m – 1 objects. This can be done in C_{m-1}^3 ways. Then, one can select two objects out of m – 1 objects (which do not include the first object) and add to each such selection the first object, originally put aside. This can be done in C_{m-1}^2 ways. By the Rule of Sum, two independent selections of three objects out of m objects, one without the first object and another one with the first object, can be done in $C_{m-1}^3 + C_{m-1}^2 = \frac{(m-1)!}{3!(m-1-3)!} + \frac{(m-1)!}{2!(m-1-2)!} = \frac{(m-1)!}{3!(m-4)!} + \frac{(m-1)!}{2!(m-3)!} = \frac{(m-1)!(m-3+3)}{3!(m-3)!} = \frac{m!}{3!(m-3)!} = C_m^3$ ways.

Input interpretation

$$\text{Table}\left[\frac{m!}{3!\,(m-3)!},\ \{m,\ \text{arithmetic progression}\ \ 3 \text{ to } 14\ \ \text{step size}\ \ 1\}\right]$$

Results

{1, 4, 10, 20, 35, 56, 84, 120, 165, 220, 286, 364}

Figure 1.13. The first 12 tetrahedral numbers generated by Wolfram Alpha.

Finally, one can see (Figure 1.14) that the number of triangles in the sequence 1, 4, 10, 20, 35, … increases by 3, 6, 10, 15 , … . The latter sequence is known as triangular numbers (partial sums of consecutive natural numbers; see also Chapter 3, Section 3.5 and Chapter 8, Section 8.4). The former sequence is known as tetrahedral (or triangular pyramidal) numbers, which are partial sums of consecutive triangular numbers. Among many contexts mentioned in the Online Encyclopedia of Integer Sequences (OEIS®), the context of Bowen's emotional system is not mentioned. One can use Wolfram Alpha (Figure 1.15) to find the formula for tetrahedral numbers

$$T(m) = \frac{m(m+1)(m+2)}{6} \tag{1.2}$$

as the partial sum of the first m triangular numbers (see also Chapter 6, Section 6.3). Alternatively, the expression $\frac{m(m+1)(m+2)}{6}$ represents the number of three-combination of $m + 2$ objects. Indeed, $C_{m+2}^{3} = \frac{(m+2)!}{3!(m+2-3)!} = \frac{m(m+1)(m+2)}{6}$. Finally, formula (1.2) shows that the product of any three consecutive natural numbers is divisible by six.

	A	B	C	D	E	F	G	H	I	J	K	L	M	N
1	1	2	3	4	5	6	7	8	9	10	11	12	13	14
2	1	3	6	10	15	21	28	36	45	55	66	78	91	105
3	1	4	10	20	35	56	84	120	165	220	286	364	455	560

Figure 1.14. From natural (row 1) to triangular (row 2) to tetrahedral (row 3) numbers.

Sum

$$\sum_{i=1}^{m} \frac{1}{2} i(i+1) = \frac{1}{6} m(m+1)(m+2)$$

Figure 1.15. Finding a formula for tetrahedral numbers using Wolfram Alpha.

Conclusion

The chapter, with reference to Thales of Miletus, highlighted triangulation as an ancient mathematical technique of using triangles in indirect measurement of heights and distances. It was shown how the origin of the word triangulation is used worldwide by research mathematicians and mathematics educators in promoting multiple methods of problem solving, including the use of technology. This demonstration was connected to the use of triangulation by sociologists as a means of integrating alternative epistemologies in the study of the same object. Vygotskian psychology ideas about the instrumental method were interpreted in terms of triangulation using several examples from school mathematics. Bowen's studies of emotional systems that used the word triangulation were discussed and mathematised to introduce basic ideas of combinatorics supported by several digital tools. The scope of this chapter is appropriate to be included in problem-solving courses for prospective teachers of multiple grades. The next chapter will show how trigonometry makes it possible to enhance the original ideas by Thales as a proxy for direct measurements.

Chapter 2

Trigonometry as a Mathematical Basis for Triangulation

2.1. Introduction

This chapter will demonstrate how the method of triangulation in mathematics stems from the similarity of right triangles and makes use of trigonometry as the extension of geometry towards the study of the relationship among angles and side lengths of the triangles. It will show how practical applications of triangulation are based on different trigonometric identities and formulas known since the 2nd century and included in the modern-day high school mathematics curriculum. The derivation of the formulas can display a variety of mathematical techniques including the use of geometric meaning of trigonometric ratios as the coordinates of points on the unit circle, the laws of Sines and Cosines, as well as the use of complex numbers connecting exponential and trigonometric functions. Focusing on more than one approach to developing a trigonometric identity can be seen as using the ideas of triangulation in trigonometry. In what follows, multiple ways, both mathematical and computational, of deriving trigonometric formulas, proving trigonometric identities, solving trigonometric equations and inequalities will be discussed. It will also be shown that methods of solving a trigonometric equation (or an algebraic equation, for that matter) may depend on the coefficients of the equation, thereby demonstrating the limitations of the method outside a specific example for which the method was successful. Just as the construction of a triangle depends on the length of segments to be used as its sides (see Chapter 6, Section 6.4), triangulation in problem solving as more than one way of making use of the conditions (explicit or hidden) of a problem may depend on the problem itself as well as on the solver (Lyness, 1968).

2.2. The Similarities of Triangles and Trigonometry in the Origins of Triangulation

This section begins with full citation from Russell (1945) who suggested (see Chapter 1, Section 1.1) that Thales "seems to have discovered how to calculate the distance of a ship in a sea from observation taken at two points on land and how to estimate the height of a pyramid from the length of its shadow" (p. 25). It is unlikely that, without knowledge of trigonometry, Thales was able to calculate an unknown distance to a ship using the known distance between two observers on land … unless, as will be shown below, the ship was looked at by each observer under 45o angle. That is why Russell put it "[Thales] seems to have discovered". In the 21st century, expectations for high school students include knowledge of trigonometric ratios as tools of solving right-angled triangles (Department for Education, 2013/2021), using "trigonometric ratios … to solve right triangles in applied problems" (Common Core State Standards, 2010, p. 77) and for instructors of high school mathematics teachers, in the context of the history of the subject matter, when "presenting mathematics as a living and evolving subject … to be aware that findings from historical research may contradict popular accounts" (Conference Board of the Mathematical Sciences, 2012, p. 61).

As shown in Figure 2.1, in trigonometry, the ratios $\frac{d}{x}$ and $\frac{d}{l-x}$ are known as $\tan\alpha$ and $\tan\beta$, respectively. The formulas, connecting the legs of a right triangle through $\tan\alpha$ or $\tan\beta$ can be used in indirect measurement of the distance from land to a ship in a sea. Figure 2.1 shows that triangulation as a mathematical technique can be seen as a process of finding an unknown location of the ship by forming a triangle the vertices of which are at the ship and at two known points defined by the distance between them; consequently, the angles of such a triangle become known. In other words, knowing angles from which the ship is seen from two points on land and the distance between the points makes it possible to find distance d along the perpendicular line that connects land and the ship. This line divides the segment between the two points from which the ship is seen under the angles α and β in the segments x and $l-x$ where l is the length of the segment connecting two points on land. That is, the known values are the length l and the angles α and β. The goal is to express d as the function of the three givens: l, α and β . When $\alpha = \beta$, the two triangles are similar so that $x = l - x$ whence $x = \frac{l}{2}$ and $d = \frac{l}{2}$.

However, when $\alpha = \beta = 45°$ we have $d = x = \frac{l}{2}$. That is how Thales might have calculated distance *d*.

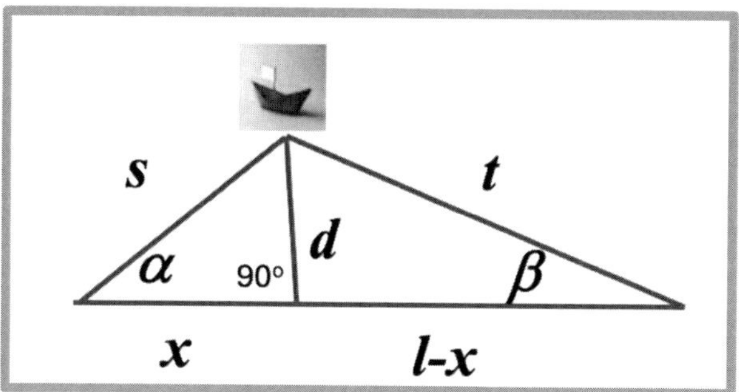

Figure 2.1. Finding distance *d* from land to a boat in a sea.

With the goal to find *d* in the general case, the first step is to express *d* through *x*, *l*, and trigonometric functions of α and β. As was mentioned above, $\frac{d}{x} = \tan\alpha, \frac{d}{l-x} = \tan\beta$ (Figure 2.1). The second step is to express *x* through *l* and trigonometric functions of α and β. We have

$$x\tan\alpha = (l-x)\tan\beta, x(\tan\alpha + \tan\beta) = l\tan\beta, x = l\frac{\tan\beta}{\tan\alpha + \tan\beta}.$$

Finally, using the relation between *d* and *x* found through the first step, the value of *d* can be obtained:

$$d = l\frac{\tan\alpha\tan\beta}{\tan\alpha+\tan\beta} = l\frac{\sin\alpha\sin\beta}{\cos\alpha\cos\beta(\frac{\sin\alpha}{\cos\alpha}+\frac{\sin\beta}{\cos\beta})} = l\frac{\sin\alpha\sin\beta}{\sin\alpha\cos\beta+\cos\alpha\sin\beta} = l\frac{\sin\alpha\sin\beta}{\sin(\alpha+\beta)}.$$

That is,

$$d = l\frac{\sin\alpha\sin\beta}{\sin(\alpha+\beta)} \tag{2.1}$$

Note that the third step was based on the formula for the sine of the sum of two angles. In the next section, this and like formulas will be developed, in the spirit of triangulation, using three different methods leading to the same result. Those formulas are sometimes called Ptolemy's identities named after Claudius Ptolemy – an Egyptian mathematician and astronomer of the 2nd century, known to be the author of the theory in which the Earth is the center of the universe – who had been using his identities when applying trigonometry to astronomy.

2.3. The Laws of Sines and Cosines in Finding Distance from Land to Ship

Formula (2.1) can be developed using an approach based on the use of two laws of trigonometry: the Law of Sines and the Law of Cosines. The Law of Sines indicates that the ratio of a side length of a triangle to the sine of the angle opposite to that side is equal to the diameter of the circumscribed circle. The very technique of using the Law of Sines in finding the side length of a triangle with known angles is called triangulation. The Law of Cosines is the generalization of the Pythagorean theorem to any triangle allowing one to find a side length of a triangle with two known sides lengths and the angle between them. These laws are studied in high school geometry as problem-solving tools that "embody the triangle congruence criteria for the cases where three pieces of information suffice to completely solve a triangle" (Common Core State Standards, 2010, p. 74).

To begin using those tools, note that according to Figure 2.1,

$$d = s \cdot \sin\alpha = t \cdot \sin\beta\ .$$

Also, according to the Law of Sines we have (Figure 2.1)

$$\frac{s}{\sin\beta} = \frac{l}{\sin(\pi-(\alpha+\beta))} = \frac{l}{\sin(\alpha+\beta)},$$

whence $s = \frac{l \cdot \sin\beta}{\sin(\alpha+\beta)}$ and $d = \frac{l \cdot \sin\alpha \cdot \sin\beta}{\sin(\alpha+\beta)}$. This confirms formula (2.1).

According to the Law of Cosines we have (Figure 2.1)

$$l^2 = s^2 + t^2 - 2s \cdot t \cdot \cos(\pi - (\alpha+\beta))$$

or $l^2 = (\frac{d}{\sin\alpha})^2 + (\frac{d}{\sin\beta})^2 + 2\frac{d}{\sin\alpha} \cdot \frac{d}{\sin\beta} \cdot \cos(\alpha+\beta)$, whence

$$(l\sin\alpha \cdot \sin\beta)^2 = d^2(\sin^2\alpha + \sin^2\beta + 2\cos(\alpha+\beta) \cdot \sin\alpha \cdot \sin\beta).$$

Input

$\sin^2(\alpha) + \sin^2(\beta) + 2\cos(\alpha + \beta)\sin(\alpha)\sin(\beta) - \sin^2(\alpha + \beta)$

Result

0

Figure 2.2. Verifying trigonometric identity (2.2) using Wolfram Alpha.

$R(\alpha, \beta) := (\sin(a))^2 + (\sin(\beta))^2 + 2\cdot\cos(\alpha + \beta)\cdot\sin(a)\cdot\sin(\beta) - (\sin(a + \beta))^2$

$R := (\alpha, \beta) \mapsto \sin(a)^2 + \sin(\beta)^2 + 2\cdot\cos(\alpha + \beta)\cdot\sin(a)\cdot\sin(\beta) - \sin(a + \beta)^2$

simplify(%)

0

Figure 2.3. Using Maple in verifying the correctness of identity (2.2).

As shown in Figure 2.2 (created by Wolfram Alpha),

$$\sin^2\alpha + \sin^2\beta + 2\cos(\alpha + \beta)\cdot\sin\alpha\cdot\sin\beta = \sin^2(\alpha + \beta). \qquad (2.2)$$

Therefore, $(l\cdot\sin\alpha\cdot\sin\beta)^2 = d^2\sin^2(\alpha + \beta)$, whence, $l\cdot\sin\alpha\cdot\sin\beta = d\cdot\sin(\alpha + \beta)$ and, confirming formula (2.1), we have, once again, $d = \frac{l\cdot\sin\alpha\cdot\sin\beta}{\sin(\alpha+\beta)}$.

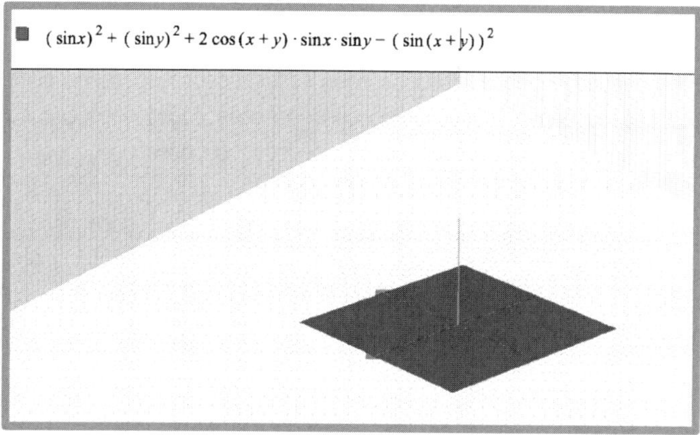

Figure 2.4. Using the Graphing Calculator in verifying the correctness of identity (2.2).

One can use Maple to triangulate the result of proving trigonometric identity (2.2). Such use of Maple is shown in Figure 2.3. Also, one can use the Graphing Calculator (Figure 2.4) to graph the difference between the left- and right-hand sides of identity (2.2) to get the coordinate plane *xOy* defined in the space by the coordinate axes *x* and *y*. The graphing component of computational triangulation "allows methods from algebra [and/or trigonometry] to be applied to geometry ... making visualization a tool for doing and understanding algebra" (Common Core State Standards, 2010, p. 74). Likewise, visualization helps secondary students in England to develop skills of seeing mathematical relationships through the combined lens of algebra and geometry (Department for Education, 2013/2021).

2.4. Deriving Ptolemy's Identities Using Two Geometric Methods

In this section, the identities/formulas for $\sin(\alpha \pm \beta)$ and $\cos(\alpha \pm \beta)$ will be derived using geometric meaning of trigonometric functions. One of such identities was used in section 2.1 for developing formula (2.1). Here, the geometric meaning of the two trigonometric functions as the coordinates of the points located on the unit circle $x^2 + y^2 = 1$ will be used. Such meaning "enables the extension of trigonometric functions to all real numbers, interpreted as radian measures of angles traversed counterclockwise around the unit circle" (Common Core State Standards, 2010, p. 71). To begin, consider the sketch of Figure 2.5. One can see that in the coordinate system (X, Y) the angles between the horizontal axis X and the segments OA and OB, respectively, are β and α. Consequently, the coordinates of the points A and B are $(\cos\beta, \sin\beta)$ and $(\cos\alpha, \sin\alpha)$, respectively. Rotating axes X and Y by angle β in the counterclockwise direction, yields new coordinates for the points A and B, namely: A = (1, 0) and B = $(\cos(\alpha - \beta), \sin(\alpha - \beta))$, respectively, in the coordinate system (X1, Y1). Because the distance between the points is the same in both coordinate systems, one can, using the distance formula between two points in the plane, equate the distances computed in the systems (X, Y) and (X1, Y1) to have

$$(\cos\beta - \cos\alpha)^2 + (\sin\beta - \sin\alpha)^2 = (1 - \cos(\alpha - \beta))^2 + \sin^2(\alpha - \beta),$$

whence

$\cos^2\beta + \sin^2\beta + \cos^2\alpha + \sin^2\alpha - 2(\cos\alpha \cdot \cos\beta + \sin\alpha \cdot \sin\beta) = 1 - 2\cos(\alpha - \beta) + \cos^2(\alpha - \beta) + \sin^2(\alpha - \beta)$, or $2 - 2(\cos\alpha \cdot \cos\beta + \sin\alpha \cdot \sin\beta) = 2 - 2\cos(\alpha - \beta)$.

That is,

$$\cos(\alpha - \beta) = \cos\alpha \cdot \cos\beta + \sin\alpha \cdot \sin\beta . \quad (2.3)$$

Replacing β by $-\beta$ in formula (2.3) yields

$\cos(\alpha + \beta) = \cos\alpha \cdot \cos(-\beta) + \sin\alpha \cdot \sin(-\beta) = \cos\alpha \cdot \cos\beta - \sin\alpha \cdot \sin\beta$.

That is,

$$\cos(\alpha + \beta) = \cos\alpha \cdot \cos\beta - \sin\alpha \cdot \sin\beta. \quad (2.4)$$

At the same time,

$$\begin{aligned} \sin(\alpha - \beta) &= \cos(\frac{\pi}{2} - (\alpha - \beta)) = \cos((\frac{\pi}{2} - \alpha) + \beta) \\ &= \cos\left(\frac{\pi}{2} - \alpha\right) \cdot \cos\beta - \sin\left(\frac{\pi}{2} - \alpha\right) \cdot \sin\beta \\ &= \sin\alpha \cdot \cos\beta - \cos\alpha \cdot \sin\beta \end{aligned}$$

That is,

$$\sin(\alpha - \beta) = \sin\alpha \cdot \cos\beta - \cos\alpha \cdot \sin\beta . \quad (2.5)$$

Replacing β by $-\beta$ in formula (2.5) yields

$$\sin(\alpha + \beta) = \sin\alpha \cdot \cos\beta + \cos\alpha \cdot \sin\beta . \quad (2.6)$$

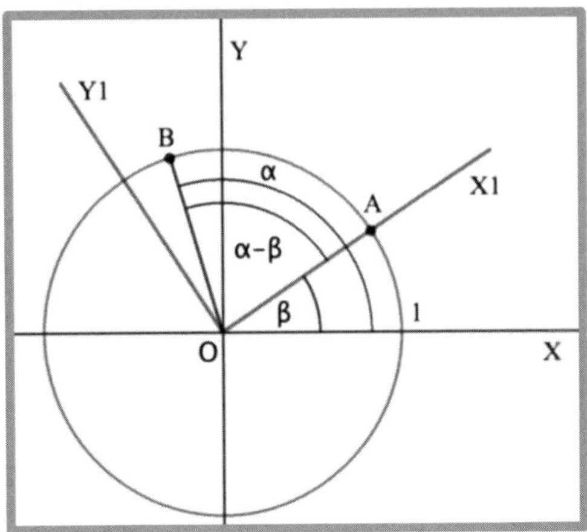

Figure 2.5. Deriving formula for $\cos(\alpha-\beta)$.

The same four formulas/identities, (2.3) – (2.6), can be obtained differently (but still geometrically). One can see (Figure 2.6) that in the coordinate system (X, Y) the angles between the horizontal axis X and the segments OA and OB, respectively, are β and α. Consequently, the coordinates of the points A and B in the coordinate system (X, Y) are $(\cos(2\pi-\beta), \sin(2\pi-\beta))$ and $(\cos\alpha, \sin\alpha)$, respectively. Rotating in the clockwise direction the axes X and Y by angle β, yields the new coordinates for the points A and B, namely, A = (1, 0) and B = $(\cos(\alpha+\beta), \sin(\alpha+\beta))$, respectively, in the coordinate system (X1, Y1). Because the distance between the points is the same in both coordinate systems, we have

$$(\cos\beta-\cos\alpha)^2+(\sin\beta+\sin\alpha)^2=(1-\cos(\alpha+\beta))^2+\sin^2(\alpha+\beta),$$

whence

$$\cos^2\beta+\sin^2\beta+\cos^2\alpha+\sin^2\alpha-2(\cos\alpha\cdot\cos\beta-\sin\alpha\cdot\sin\beta)=$$
$$1-2\cos(\alpha+\beta)+\cos^2(\alpha+\beta)+\sin^2(\alpha+\beta)=2-2\cos(\alpha+\beta).$$

That is, $\cos(\alpha+\beta)=\cos\alpha\cdot\cos\beta-\sin\alpha\cdot\sin\beta$. Replacing β by $-\beta$ in the last formula yields $\cos(\alpha-\beta)=\cos\alpha\cdot\cos\beta+\sin\alpha\cdot\sin\beta$. The last two formulas are exactly formulas (2.4) and (2.3). From here, formulas (2.5) and (2.6) can be obtained.

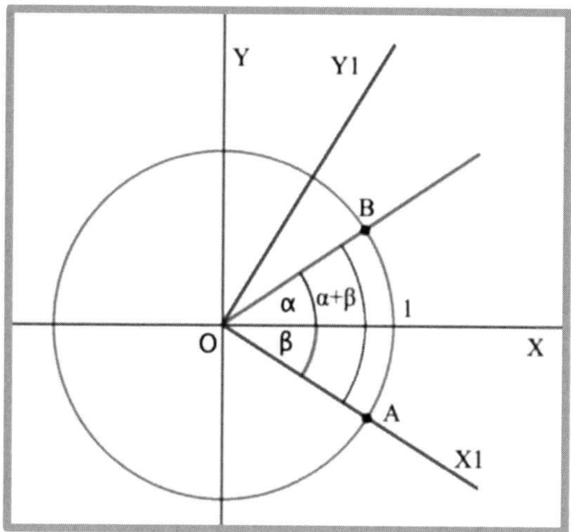

Figure 2.6. Deriving formula for $\cos(\alpha + \beta)$.

2.5. Deriving Ptolemy's Identities Using Complex Numbers

The third method to develop formulas/identities (2.3) – (2.6) is based on the use of complex numbers, "another extension of number, when the real numbers are augmented by imaginary numbers … [so that] the commutative, associative, and distributive properties and their new meanings are consistent with their previous meanings" (Common Core State Standards, 2010, p. 58). With this in mind, formulas for trigonometric functions can be used. One of the most famous mathematical formulas developed around 1740 by a Swiss mathematician Leonhard Euler (1707-1783), the father of all modern mathematics,

$$e^{xi} = \cos x + i \sin x \tag{2.7}$$

connects the exponential and trigonometric functions, demonstrating one of "the surprising relations between the elementary transcendentals" (Klein, 2016, p. 81). It follows from formula (2.7) that

$e^{\alpha i} = \cos\alpha + i\sin\alpha$ and $e^{-\alpha i} = \cos\alpha - i\sin\alpha$, whence $\cos\alpha = \frac{e^{\alpha i}+e^{-\alpha i}}{2}$ and $\sin\alpha = \frac{e^{\alpha i}-e^{-\alpha i}}{2i}$. Using the last two formulas, we have

$$\sin(\alpha+\beta)=\frac{e^{(\alpha+\beta)i}-e^{-(\alpha+\beta)i}}{2i}=\frac{e^{\alpha i}e^{\beta i}-e^{-\alpha i}e^{-\beta i}}{2i},$$

$$\sin\alpha\cdot\cos\beta=\frac{e^{\alpha i}-e^{-\alpha i}}{2i}\cdot\frac{e^{\beta i}+e^{-\beta i}}{2}=\frac{e^{(\alpha+\beta)i}+e^{(\alpha-\beta)i}-e^{(\alpha-\beta)i}-e^{-(\alpha+\beta)i}}{4i},$$

$$\cos\alpha\cdot\sin\beta=\frac{e^{\alpha i}+e^{-\alpha i}}{2}\cdot\frac{e^{\beta i}-e^{-\beta i}}{2i}=\frac{e^{(\alpha+\beta)i}-e^{(\alpha-\beta)i}+e^{-(\alpha-\beta)i}-e^{-(\alpha+\beta)i}}{4i}.$$

Therefore,

$$\sin\alpha\cdot\cos\beta+\cos\alpha\cdot\sin\beta=\frac{2e^{(\alpha+\beta)i}-2e^{-(\alpha+\beta)i}}{4i}=\frac{e^{(\alpha+\beta)i}-e^{-(\alpha+\beta)i}}{2i}$$
$$=\sin(\alpha+\beta)$$

and formula (2.6) is confirmed. Replacing β by $-\beta$ yields formula (2.5). Note that the approach based on the use of Euler's formulas for trigonometric functions is purely algebraic and it can be used for complex arguments of trigonometric functions when their geometric meaning does not make sense. In terms of mathematics teacher education, opportunities of dealing with "complex numbers and trigonometry together can prepare teachers with a solid foundation in both areas, and it can help them make sense of some seemingly disconnected terrains in upper-level high school mathematics … [because through] tight connection between complex numbers and trigonometry many trigonometric identities can be viewed as algebraic identities in **C** [the set of complex numbers]" (Conference Board of the Mathematical Sciences, 2012, p. 64).

2.6. Triangulation as Solving Trigonometric Equations Using Different Techniques

Consider the equation

$$\cos^2 2x+3=(\sin^2 x-3)^2 \qquad (2.8)$$

Three methods of solving equation (2.8) will be considered under the umbrella of triangulation demonstrating how different methods yield the same result (answer). The equation was chosen to allow for multiple techniques leading to the same symbolic form of the answer. More often, different problem-solving techniques in trigonometry (used as a triangulation of results) yield different symbolic forms of answers. In such cases, computational triangulation of the second order is required to demonstrate that symbolically different forms of answers are numerically identical.

The first method of solving equation (2.8) is to reduce it to an equation about $\sin^2 x$ by using the formula $\cos 2x = 1 - 2\sin^2 x$ which follows from formula (2.4) by representing $\cos 2x = \cos(x + x) = \cos x \cdot \cos x - \sin x \cdot \sin x = \cos^2 x - \sin^2 x$ and then using the formula $\cos^2 x = 1 - \sin^2 x$ to get $\cos 2x = 1 - 2\sin^2 x$. Setting $y = \sin^2 x$, $0 \le y \le 1$, equation (2.8) turns into the equation $(1 - 2y)^2 + 3 = (y - 3)^2$, whence $3y^2 + 2y - 5 = 0$. The last quadratic equation has two roots: $y = 1, y = -\frac{5}{3}$. Rejecting the second (negative) root results in the equation $\sin^2 x = 1$ or $\sin x = \pm 1$, whence $x = \pm\frac{\pi}{2} + \pi n, n = 0, \pm 1, \pm 2, \ldots$.

The second method of solving equation (2.8) is to reduce it to an equation about $\cos^2 x$ by using the formulas $\cos 2x = 2\cos^2 x - 1$ and $\sin^2 x = 1 - \cos^2 x$. Setting $y = \cos^2 x$, $0 \le y \le 1$, equation (2.8) turns into the equation $(2y - 1)^2 + 3 = (2 + y)^2$, whence $3y^2 - 8y = 0$. The last quadratic equation has two roots: $y = 0, y = \frac{8}{3}$. Rejecting the second (greater than one) root results in the equation $\cos^2 x = 0$ which is equivalent to $\sin^2 x = 1$, whence, once again, $\sin x = \pm 1$, and $x = \pm\frac{\pi}{2} + \pi n, n = 0, \pm 1, \pm 2, \ldots$.

The third method of solving equation (2.8) is to reduce it to an equation about $\cos 2x$ by using the formula $\cos^2 x = \frac{1+\cos 2x}{2}$ which is a slight variation of the formula $\cos 2x = 2\cos^2 x - 1$ used in the second method. Setting $y = \cos 2x$, $|y| \le 1$, equation (2.8) turns into the equation $y^2 + 3 = (\frac{y+5}{2})^2$, whence $3y^2 - 10y - 13 = 0$. The last quadratic equation has two roots: $y = -1, y = \frac{13}{3}$. Rejecting the second (greater than one) root yields $\cos 2x = -1$ or $1 - 2\sin^2 x = -1$, whence $\sin^2 x = 1$ yielding, once again, $\sin x = \pm 1$, and $x = \pm\frac{\pi}{2} + \pi n, n = 0, \pm 1, \pm 2, \ldots$.

Remark 2.1. Had we replaced the number 3 by the number 2 in equation (2.8), the same three methods would give symbolically different answers in each case, requiring one, as was mentioned above, to use computational triangulation of the second order to demonstrate that different symbolic representations with different arc functions are, in fact, numerically identical. More specifically, the first method (using the formula $\cos 2x = 1 -$

$2\sin^2 x$) yields $x = \pm\sin^{-1}\frac{1}{\sqrt[4]{3}} + \pi n$; the second method (using the formula $\cos 2x = 2\cos^2 x - 1$) yields $x = \pm\cos^{-1}(\pm\sqrt{\frac{3-\sqrt{3}}{3}}) + 2\pi n$; the third method (using the formula $\cos^2 x = \frac{1+\cos 2x}{2}$) yields $x = \frac{\pm\cos^{-1}(\frac{3-2\sqrt{3}}{3})}{2} + \pi n$.

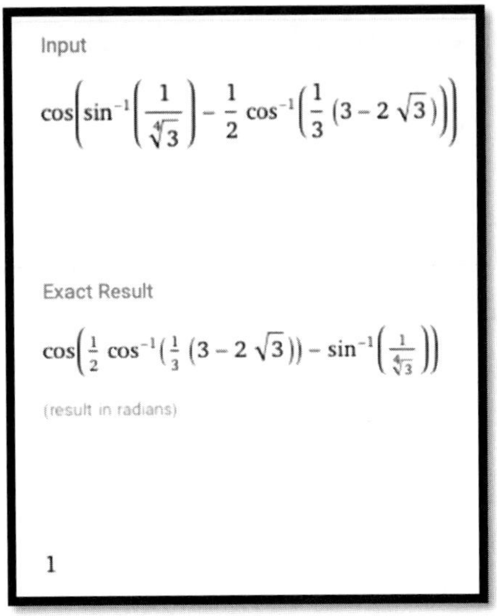

Input

$$\cos\left(\sin^{-1}\left(\frac{1}{\sqrt[4]{3}}\right) - \frac{1}{2}\cos^{-1}\left(\frac{1}{3}(3-2\sqrt{3})\right)\right)$$

Exact Result

$$\cos\left(\frac{1}{2}\cos^{-1}\left(\frac{1}{3}(3-2\sqrt{3})\right) - \sin^{-1}\left(\frac{1}{\sqrt[4]{3}}\right)\right)$$

(result in radians)

1

Figure 2.7. Wolfram Alpha provides computational triangulation of the second order.

Figure 2.7 shows how Wolfram Alpha provides computational triangulation of the second order in proving the identity of answers obtained by the first and the third methods as it demonstrates that the difference $\sin^{-1}\frac{1}{\sqrt[4]{3}} - \frac{\cos^{-1}(\frac{3-2\sqrt{3}}{3})}{2}$ is equal to zero because its cosine is equal to one. Indeed, both the minuend and the subtrahend are positive and smaller than $\pi/2$ and, therefore, their difference may not be equal to $\pm 2\pi$, another two angles in the range $|x| \leq 2\pi$, besides zero, the cosine of which is equal to one.

Remark 2.2. One can check to see that replacing 3 by 2 in equation (2.8) would allow for the fourth method of reducing the so modified equation to a quadratic equation about $\tan^2 x$, because, in that case, $\cos x \neq 0$ in the equation $\cos^2 2x + 2 = (\sin^2 x - 2)^2$. However, $\cos x = 0$ does satisfy equation (2.8) making such reduction impossible for it makes $\tan x$

undefinable. Indeed, when $x = \frac{\pi}{2}$ we have $\cos\frac{\pi}{2} = 0$ and $\cos^2 \pi + 3 = (\sin^2 \frac{\pi}{2} - 3)^2$ in equation (2.8). That is, the method of triangulation in a trigonometric equation may depend on the domain of definition of the unknowns involved. Furthermore, a method of solving a trigonometric equation may depend on coefficients of that equation. (See also the discussion about solving equations (3.1) and (3.2) in Chapter 3, Section 3.3).

2.7. Triangulation as Solving Trigonometric Inequality Using Different Techniques

Consider the inequality

$$\cos 2x > \cos x - \sin x \tag{2.9}$$

which has to be solved for $x \in (0, 2\pi)$, where 2π is a common period of the three functions $y = \sin x, y = \cos x, y = \cos 2x$. Several problem-solving techniques will be discussed as a combination of the direct use of software along with the joint use of trigonometric machinery and computer graphing. The first approach is to outsource solving inequality (2.9) to Wolfram Alpha (Figure 2.8). Alternatively, one can graph the inequality using the Graphing Calculator (Figure 2.9). By graphing the vertical lines $x = \frac{\pi}{4}, x = \frac{\pi}{2}, x = \frac{5\pi}{4}$, one can confirm visually the result generated by Wolfram Alpha: the graph of the function $y = \cos 2x$ resides above the graph of the function $y = \cos x - \sin x$ on the intervals $(0, \frac{\pi}{4})$ and $(\frac{\pi}{2}, \frac{5\pi}{4})$. The same behavior is shown in Figure 2.10 and Figure 2.11 (left).

The second approach is to modify inequality (2.9) to the form of the product of two trigonometric expressions connected to zero through an inequality sign. This approach reduces the so modified inequality to two systems of basic trigonometric inequalities to allow for solution to be found by computer graphing. The third approach is similar to the second one with difference in trigonometric techniques used.

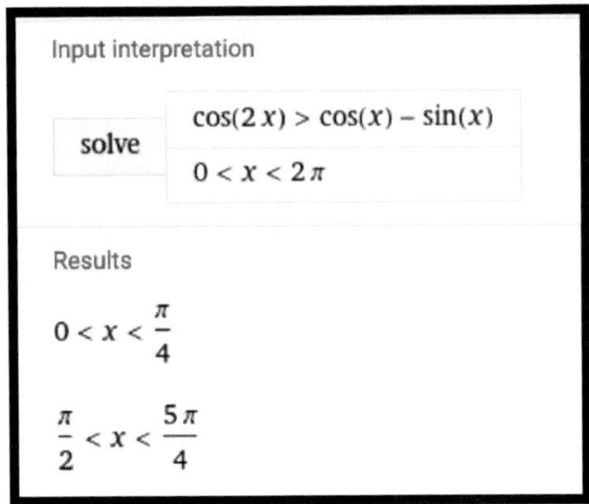
Input interpretation

solve	$\cos(2x) > \cos(x) - \sin(x)$
	$0 < x < 2\pi$

Results

$0 < x < \frac{\pi}{4}$

$\frac{\pi}{2} < x < \frac{5\pi}{4}$

Figure 2.8. Solving inequality (2.9) by Wolfram Alpha.

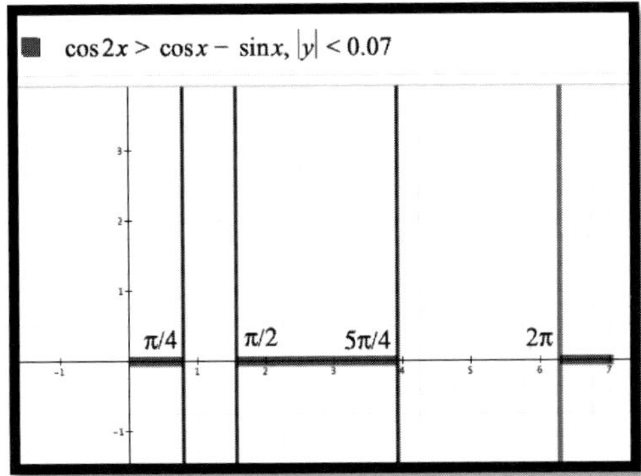

Figure 2.9. Solving inequality (2.9) using the Graphing Calculator.

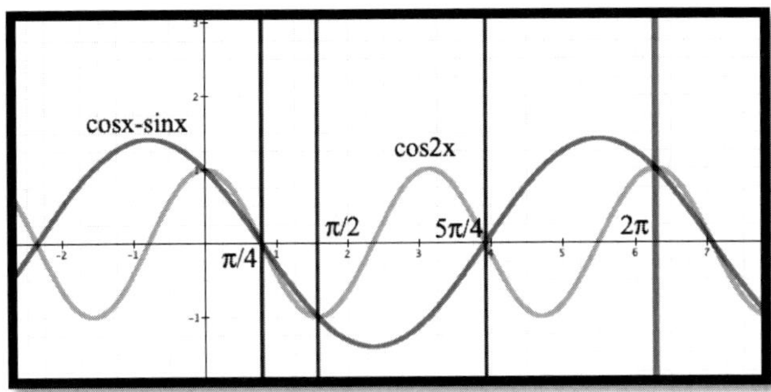

Figure 2.10. Seeing where the left- and right-hand sides satisfy inequality (2.9).

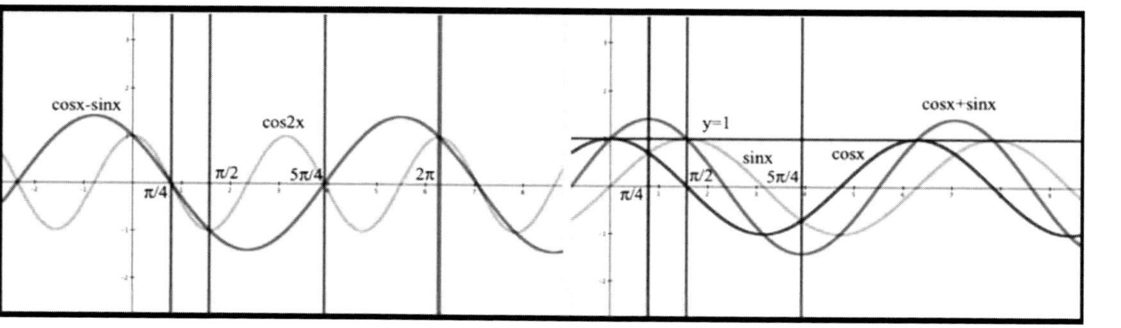

Figure 2.11. Graphing inequality (2.9) – left, and systems (2.10) – right.

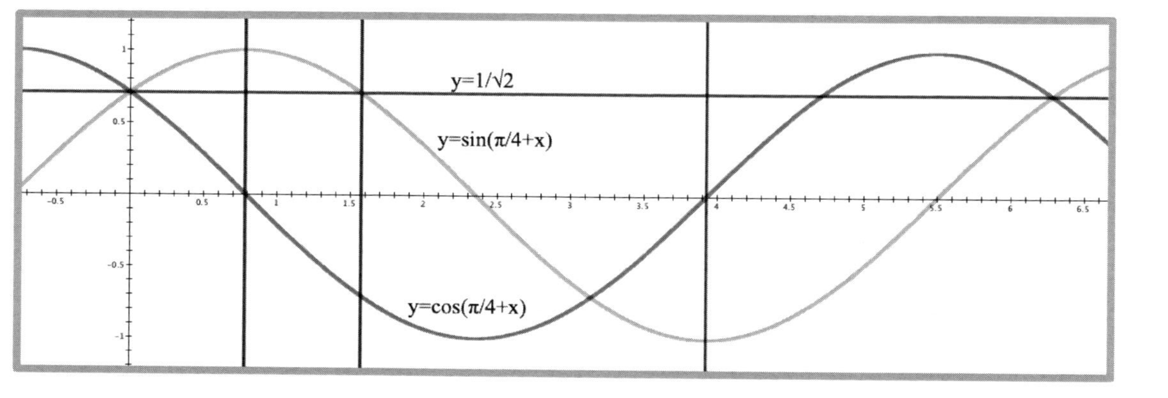

Figure 2.12. Graphing $y = \sin\left(\frac{\pi}{4} + x\right), y = \cos\left(\frac{\pi}{4} + x\right), y = \frac{1}{\sqrt{2}}, x = \frac{\pi}{4}, x = \frac{\pi}{2}, x = \frac{5\pi}{4}$.

Using the second approach, note that due to the formula $\cos 2x = \cos^2 x - \sin^2 x$, inequality (2.9) can be rewritten in the form $(\cos x - \sin x)(\cos x + \sin x) - (\cos x - \sin x) > 0$ or $(\cos x - \sin x)(\cos x + \sin x - 1) > 0$. The product of two factors is positive when both factors are either positive or negative. Therefore, the last inequality, along with inequality (2.9), is equivalent to the unity of two systems of inequalities

$$\begin{cases} \cos x - \sin x > 0 \\ \cos x + \sin x > 1 \end{cases}, \begin{cases} \cos x - \sin x < 0 \\ \cos x + \sin x < 1 \end{cases}$$

or

$$\begin{cases} \cos x > \sin x \\ \cos x + \sin x > 1 \end{cases}, \begin{cases} \cos x < \sin x \\ \cos x + \sin x < 1 \end{cases}. \tag{2.10}$$

The last two systems of inequalities can be solved through graphing the functions $y = \cos x, y = \sin x, y = \cos x + \sin x$ (Figure 2.11, right) to see where the required inequalities hold true. By graphing the vertical lines $x = \frac{\pi}{4}, x = \frac{\pi}{2}, x = \frac{5\pi}{4}$, one can see that for $0 < x < \frac{\pi}{4}$ the inequalities from the first system of (2.10) hold true and for $\frac{\pi}{2} < x < \frac{5\pi}{4}$ the inequalities from the second system of (2.10) hold true.

The third method of solving inequality (2.9) is to replace its left- and right-hand sides as follows: $\cos 2x = \sin\left(\frac{\pi}{2} + 2x\right) = 2\sin(\frac{\pi}{4} + x)\cos(\frac{\pi}{4} + x)$; $\cos x - \sin x = \sqrt{2}\cos(\frac{\pi}{4} + x)$.

Inequality (2.9) can be re-written in the form

$$2\sin(\frac{\pi}{4} + x)\cos(\frac{\pi}{4} + x) - \sqrt{2}\cos\left(\frac{\pi}{4} + x\right) > 0$$

or

$$2\cos\left(\frac{\pi}{4} + x\right)[\sin\left(\frac{\pi}{4} + x\right) - \frac{1}{\sqrt{2}}] > 0. \tag{2.11}$$

Inequality (2.11) is equivalent to the unity of two systems of inequalities

$$\begin{cases}\cos\left(\frac{\pi}{4}+x\right)>0\\ \sin\left(\frac{\pi}{4}+x\right)>\frac{1}{\sqrt{2}}\end{cases}, \begin{cases}\cos\left(\frac{\pi}{4}+x\right)<0\\ \sin\left(\frac{\pi}{4}+x\right)<\frac{1}{\sqrt{2}}\end{cases}. \qquad (2.12)$$

Figure 2.12 shows that the inequalities of the first system (2.12) hold true for $0 < x < \frac{\pi}{4}$ and the inequalities of the second system (2.12) hold true for $\frac{\pi}{2} < x < \frac{5\pi}{4}$ thereby confirming the results obtained by other methods.

In that way, using multiple problem-solving techniques in the context of trigonometric equations and inequalities may be considered through the lens "of bringing mathematical coherence to high school mathematics … replacing much of the special-purpose paraphernalia that clutters many high school programs, such as the various mnemonics for formulas in trigonometry" (Conference Board of the Mathematical Sciences, 2012, p. 67). One of the author's students, a teacher candidate, put it simply as follows: "*There is usually more than one way to solve a problem. Therefore, teachers should fully understand the concepts and understand the different methods to solve the problem. From my experience, some teachers are more about the method of learning instead of the concept that is being taught. I think we cause students to lose interest in math when we expect them to complete it a certain way.*" Another teacher candidate explains that the preference for only one way is because "*a teacher may only know one way, and if that's all they know, how are they going to teach any other way. And this could hurt the student, later on, placing them even farther behind from a child that had a teacher that was well trained in math.*" These comments of teacher candidates implicitly support the value of computation triangulation as a tool of pedagogy of the modern-day mathematics teacher education.

Conclusion

The chapter demonstrated the use of triangulation as a method of multiple solution strategies in secondary school trigonometry. This made it possible to enhance the ancient technique of indirect measurement by means of the laws of Sines, Cosines and Ptolemy's identities which, in the spirit of triangulation, were derived by different methods. Supported by more than one digital instrument, the ideas of computational triangulation were integrated with solving trigonometric equations and inequalities. The scope of this chapter is appropriate to be included in problem-solving courses for prospective teachers of multiple grades. The next chapter will provide practical and theoretical ideas about using more than one digital tool in triangulating mathematical problem solving.

Chapter 3

Computational Triangulation, Advanced Conceptual Understanding and TITE Framework

3.1 Introduction

Whereas the history of the term triangulation suggests that its nature is "unsettling and unruly" (Denzin, 2007, p. 5079), the idea behind using triangulation outside of mathematics as its cradle was to assist qualitative social science researchers with rigorous methodologies. Some mathematics teachers, with conservative (sometime resistant to be changed) beliefs developed during their own learning of the subject matter (Cooney, 2001; Swan, 2007; Liljedahl, et al., 2012), might suggest that solving a problem in more than one way causes anxiety and confusion among students, especially the younger ones, thereby diminishing teachers' control of the classroom. Nonetheless, triangulation in mathematics education, as more than one way of arriving at the answer, provides more rigor to the process of problem solving by connecting different concepts, using alternative techniques and, in the digital era, performing computational experiments. Consistent with sociology research, when mathematics educators study teacher candidates' opinion about triangulation in problem solving, different data sources are used such as completed assignments, classroom presentations, personal communications, solicited reflections, and discussion boards.

The credibility of results in mathematical problem solving with experimental (not necessarily digital) assistance was described in pre-digital era by Freudenthal (1978) as follows: "It is independency of new experiments that enhances credibility … [for] repeating does not create new evidence, which in fact is successfully aspired to by independent experiments" (pp. 193–194). Nowadays, whereas "engineering artifacts are unfailingly reliable ... in the realm of computers, unreliability sometimes seems to be the norm" (Harrison, 2008, p. 1400), it is the reality of this 'norm' that calls for triangulation in the use of more than one computer program when checking the results of computational problem solving.

3.2. Towards the Credibility of Problem Solving Through Computational Triangulation

In the digital era, teacher candidates "should become familiar with various software programs and technology platforms, learning how to use them to analyze data, to reduce computational overload, to build computational models of mathematical objects, and to perform mathematical experiments" (Conference Board of the Mathematical Sciences, 2012, p. 57). However, when symbolic computations are outsourced to software, the accuracy of the result is hanging on the accuracy of an algorithm involved, including the accuracy of the programming code used. For example, typing in the context of Maple the product $(\sqrt{n}+1)(\sqrt{n}-1)$, expecting it to be expanded, yields $\sqrt{n(\sqrt{n}-1)}+1$, which, however, is not correct as $(\sqrt{n}+1)(\sqrt{n}-1)=n-1$. The error here is due to the missing multiplication sign between the factors (parentheses) in the code. One may try to find out Maple's understanding of the expression $(\sqrt{n}+1)(\sqrt{n}-1)$ that yields $\sqrt{n(\sqrt{n}-1)}+1$. Note that changing the order of the factors and typing the product as $(\sqrt{n}-1)(\sqrt{n}+1)$ yield $\sqrt{n(\sqrt{n}+1)}-1$. It appears that the first square root is applied to the product $n(\sqrt{n}\pm 1)$ and then either 1 or -1 is added to this large (external) square root. Replacing the square roots by the cube roots makes things even more confusing if the erroneous result has to be interpreted. Yet, typing $(\sqrt{n}+1)*(\sqrt{n}-1)$ with the asterisk between two factors as the legit multiplication sign in Maple is not enough as it does not produce the desired result, $n-1$, either. What is needed is to follow with the command "expand (%)" (alternatively, "collect (%, n)") which would yield the correct result (the percentage symbol in the Maple language means "the latter"). At the same time, typing in the input box of Wolfram Alpha $(\sqrt{n}+1)(\sqrt{n}-1)$ does provide the correct result, yet along with information that might confuse users if they do not understand what kind of result one should select from the information provided, following a simple algebraic multiplication of two binomials.

Another issue to be aware of when dealing with multiple tools is the notation used. Consider the case of the greatest integer function which, when applied to x, returns the greatest integer smaller than or equal to x; e.g., when applied to 3.3 or to 3 returns 3 in either case. In Wolfram Alpha (see Chapter 6, Section 6.6), typing INT (3.3) or Floor (3.3) yields 3, typing int (3.3) yields 3.3*x* + constant as the value of the indefinite integral $\int 3.3dx$. In Maple, typing floor (3.3) yields 3, typing Floor (3.3) yields nothing, and typing int (3.3) requires more information treating this notation as integration. In the Graphing Calculator, only the notation floor (lower case f) is understood correctly as the greatest integer function (see Chapter 6, Section 6.6). The spreadsheet understands the input = int (3.3) as 3, yet it requires another argument for the input = floor (3.3) so that the input = floor (3.3,1) yields 3 and the input = floor (3.3, 2) yields 2 as it rounds 3.3 down to either the nearest multiple of 1 or 2, respectively (and therefore the input =floor (3.3,4) yields 0). The above-mentioned information about code and notation is correct by the time of writing this book; nonetheless, it could be revised by software developers without necessarily notifying users.

The issue of the intricacy of digital symbolic computations was further recognized by Conole & Dyke (2004) who noted that in the digital age, with an easy access to the Internet, one has to learn how to manage the abundance of information provided and how to use mathematical notation when coding might be changed as a new version of software is published. Indeed, in the words of one teacher candidate, the author's student, "*with the use of technology, students have much more information that is accessible and are constantly learning and evolving. With this comes a greater diversity of problem solving.*" The candidate implicitly points at the variety of information the Internet provides and is explicit in their belief that the diversity of problem solving is positively affected by students' use of technology. However, even solving simple mathematical problems with technology requires certain level of mathematical sophistication and accuracy in computational thinking (Wing, 2006) in order to validate the result obtained through computational triangulation using several digital instruments. Furthermore, when using technology in problem solving one should begin to think about a problem in hand with some initial understanding of how its solution might look like (Arnheim, 1969) and why the solution does not look like the one expected. For example, by typing in the input box of the Graphing Calculator the product $(\sqrt{n}+1)(\sqrt{n}-1)$ does not result in any graph (as n is treated as a slider-controlled parameter), by typing $y=(\sqrt{n}+1)(\sqrt{n}-1)$ results in the line y = constant (Figure 3.1, left; $y = -1$ when the slider value of n is zero), and by typing $y=(\sqrt{x}+1)(\sqrt{x}-1)$ results in the slant line (Figure 3.1, right; as $\sqrt{x}$ is defined for $x \geq 0$) stemming from the point (0, -1). This kind of understanding may be developed through interaction with peers and the instructor.

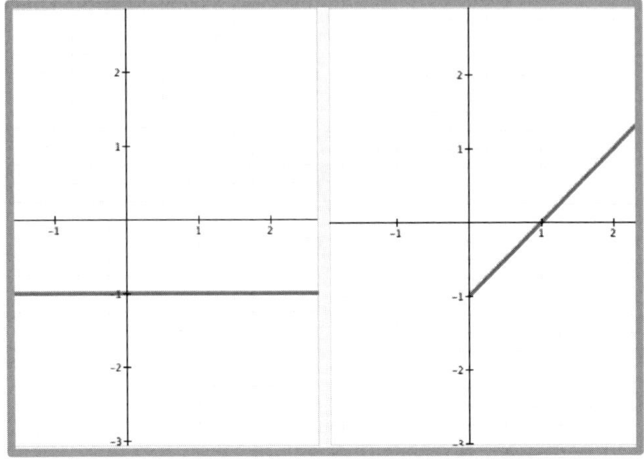

Figure 3.1. Images of $(\sqrt{n}+1)(\sqrt{n}-1)$ – left and $(\sqrt{x}+1)(\sqrt{x}-1)$ – right.

3.3. Advanced Conceptual Understanding as the Core of Theoretical Triangulation

As was mentioned elsewhere (Abramovich, 2015), one can distinguish between two levels of conceptual understanding used in mathematical problem solving – basic conceptual understanding (BCU), and advanced conceptual understanding (ACU). Both BCU and ACU are important tools of computational triangulation. The purpose of using BCU is to activate problem solving. ACU can serve at least two purposes: complete problem solving by advancing BCU and pose a new problem by reflecting on the one already solved. The process of triangulation, including computational triangulation, in mathematics education and within its teacher education component, involves continuous interaction between problem solving and problem posing. Through this process, ACU at one level can be used as a BCU at another, higher level of mathematical thinking, which then leads to a new ACU. Put another way, "teachers need to know how the mathematics they teach is connected with that of prior and later grades" (Conference Board of the Mathematical Sciences, 2012, p. 1). A teacher candidate, one of the author's students explained that such knowledge would benefit students: "*We can best meet the needs of all of the students in our class if we are aware of how the mathematics concepts connect across grade levels. If we know where our students' knowledge falls in relation to the Common Core State Standards, we can better design instruction to meet their needs and ensure that everyone is on track with their learning.*" In what follows, it is assumed that the appropriate combination of BCU and ACU is required in order to carry out triangulation as multiple ways of solving a problem that converge to the equivalent representations of the results where the equivalence of different symbolic forms requires using computational triangulation of the second order.

According to Denzin (1970), one of the types of sociological triangulations is theoretical triangulation "that few investigators achieve" (p. 303) because they often deal with many contradictory propositions. In mathematics, there are many contrasting problem-solving techniques, the appropriate use of which requires ACU. For example, the technique of dividing both sides of an equation by a variable expression works in one case and does not work in another case. To clarify, consider two cases from trigonometry. The first case is to divide both sides of the equation (see also Remark 2.2, Chapter 2)

$$\sin x \cos x - \sin^2 x = 0 \tag{3.1}$$

by $\cos^2 x$ to get the equation $\tan x - \tan^2 x = 0$ whence $\tan x = 0$ or $\tan x = 1$ (i.e., $x = \pi n$ or $x = \frac{\pi}{4} + \pi n, n = 0, \pm 1, \pm 2, \ldots$). The second case is to divide both sides of the equation

$$\sin x \cos x - \cos^2 x = 0 \tag{3.2}$$

by $\cos^2 x$ to get the equation $\tan x = 1$. One can recognize that in the first case $\tan x = 0$ is a solution and in the second case it is not (as $\sin \pi n = 0$ and $\cos \pi n = \pm 1, n = 0, \pm 1, \pm 2, \ldots$). In the case of equation (3.1), the division by $\cos^2 x$ preserves the equivalence of solutions of the original and the resulting equations; in the case of equation (3.2), the equivalence of solutions is violated due to the division by $\cos^2 x$ and the solutions satisfying the equation $\cos x = 0$ (i.e., $x = \frac{\pi}{2} + \pi n, n = 0, \pm 1, \pm 2, \ldots$) are lost. A similar outcome can be observed when dividing both sides of (3.1) and (3.2) by $\sin^2 x$. The loss of the solutions of equation (3.2) is shown graphically in Figure 3.2 demonstrating the difference in the number of intersections of the x-axis by the graphs of the functions $y = \sin x \cdot \cos x - \cos^2 x$ and $y = \tan x - 1$ in the interval of length 2π – the common period of the two functions; the latter function does not share those intersections with the graph of the function $y = \cos x$ at the points $x = \frac{\pi}{2}$ and $x = \frac{3\pi}{2}$. Yet jointly, the graphs of the functions $y = \cos x$ and $y = \tan x - 1$ intersect the x-axis within $[0, 2\pi]$ at the same points as the graph of the function $y = \sin x \cdot \cos x - \cos^2 x$ (i.e., when $x \in \{\frac{\pi}{4}, \frac{\pi}{2}, \frac{5\pi}{4}, \frac{3\pi}{2}\}$).

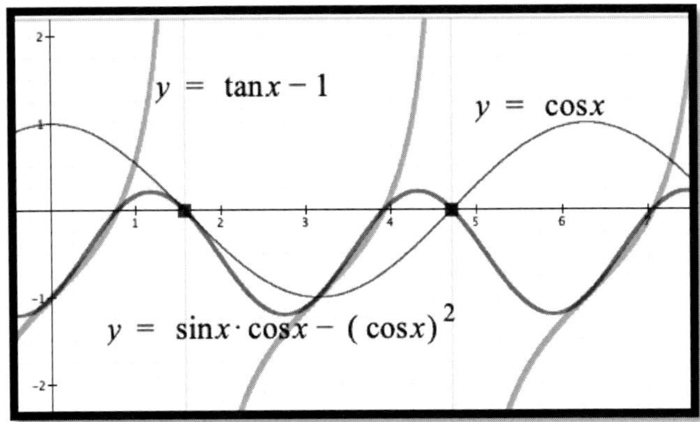

Figure 3.2. The case of equation (3.2).

This contradiction in the use of seemingly similar strategies can be explained in terms of the Einstellung effect (Luchins, 1942) or doubled experience (Vygotsky, 1999). The study of Einstellung effect, described as one's tendency to use previously learned workable strategy in situations that either can be resolved more efficiently through a different approach or to which the strategy is not applicable at all, has been carried out in the context of the so-called water jar experiments. According to Levitt (1956), originally, a water jar test was administered by Zenner and Dunker in Germany in the 1920's to explore whether one is capable to bypass the mindset developed through practicing a particular problem-solving technique and, in doing so, to demonstrate insight (Dunker, 1945) and/or skill of productive thinking (Wertheimer, 1959). In the words of Luchins, who, according to (Bilalić et al., 2010), was the first to carry out water jar experiments in the United States, "*Einstellung – habilitation – creates a mechanized state of mind, a blind attitude toward problems; one does not look at the problem on its own merits but is led by a mechanical application of a used method*" (Luchins, 1942, p. 15, italics in the original).

The concept of doubled experience, introduced by Vygotsky (1999) with reference to Marx (1982) is a similar psychological phenomenon when the final result of a professional activity (in our case, solving a trigonometric equation) repeats what earlier was present in one's mind when being engaged in another, seemingly related, activity. That is, one repeats the problem-solving technique that was successful for a different but somewhat similar equation. Both concepts, Einstellung effect and doubled experience, can describe the fixed mind set (Einstellung) which, due to successful dividing both sides of (3.1) by $\cos^2 x$ because the equation $\cos x = 0$ does not provide a solution to (3.1), transfers the strategy to the case of dividing both sides of (3.2) by $\cos^2 x$ which leads to the loss of solutions because $\cos x = 0$ is the solution of (3.2). Computational triangulation, and triangulation approach, in general, have potential to reduce the effect of Einstellung in problem solving by emphasizing the diversity of approaches, either through the variation of software or mathematical methods.

Likewise, one can show through a conceptual shortcut (Canobi, 2005; Kuo et al., 2013; Abramovich, 2024) that $x = 1$ is the only root of an irrational equation

$$\sqrt{2-x} = x\,, \tag{3.3}$$

a strategy that allows one to come to this conclusion without squaring or graphing both sides of equation (3.3). Indeed, as $0 \le x \le 2$, two cases need to be considered: $0 \le x < 1$ and $1 \le x \le 2$. In the former case, i.e., when $x < 1$, we have a contradiction as $1 < 2 - x$ and, therefore, $x = \sqrt{2-x} > 1$. In the latter case, i.e., when $x \ge 1$, we also have a contradiction (unless $x = 1$ when equation (3.3) turns into the equality $\sqrt{1} = 1$) as $1 \ge 2 - x$ and, therefore, $x = \sqrt{2-x} \le 1$. At the same time, squaring and graphing both sides of equation (3.3) can be used as a triangulation between the three methods – establishing contradictions through a conceptual shortcut, squaring both sides of equation (3.3) and solving the resulting quadratic equation for $x > 0$, and constructing the graphs of the functions $y = \sqrt{2-x}$ and $y = x$ to observe their single intersection at $x = 1$.

At the same time, the equation $\sqrt{2-x} = |x|$ has two roots: $x = 1$ and $x = -2$. The negative root can also be found through the process of triangulation: without squaring, by squaring, and by graphing. When $x < 0$ the last equation turns into $\sqrt{2-x} = -x$ or $\sqrt{2+t} = t$, where $t = -x >$

0. From transforming the difference $\sqrt{2+t}-t=\frac{(\sqrt{2+t}-t)(\sqrt{2+t}+t)}{\sqrt{2+t}+t}=\frac{2+t-t^2}{\sqrt{2+t}+t}=-\frac{(t+1)(t-2)}{\sqrt{2+t}+t}$ it follows that $\sqrt{2+t}<t$ when $t>2$, $\sqrt{2+t}>t$ when $0<t<2$, and $\sqrt{2+t}=t$ when $t=2$; i.e., $x=-2$. Each method, despite a belief by McFee (1992) that "there cannot be triangulation *between* methods" (p. 217, italics in the original), addresses the same issue, has its own value, and highlights its specific advantage and possible drawback. Developing such ACU of mathematical problem solving when working with teacher candidates, just as theoretical triangulation in sociology (Denzin, 1970), is, indeed, not always possible to achieve.

3.4. Using Multiple Graphing Tools as Data Sources Triangulation

One method of exploring the behavior of functions is through the construction of their graphs. Nowadays, plotting graphs of functions is mostly digital. Using different graphing tools may be seen as triangulation within a method. Consider the case of graphing the function $y=\frac{x^2-1}{|x|-1}$. Figure 3.3 shows (from left to right) the use of three digital tools – the Graphing Calculator, Wolfram Alpha, and Maple in graphing this function. One can see that the first two tools do not mark the points at $x=\pm1$ as not belonging to the graph of the function $y=\frac{x^2-1}{|x|-1}$. Only Maple shows the accurate graph, with a hole on each branch. This technological inconsistency among the tools suggests at least two things: the value of ACU of mathematics behind the construction of the graphs of functions that requires, as was mentioned above, some initial understanding of how the final result should look like, and the asset of the idea of computational triangulation because using more than one digital tool enhances the credibility of graphing and, more generally, of mathematical problem solving (Freudenthal, 1978).

At the same time, some level of ACU can be achieved through computational triangulation without advanced knowledge of the final result, including the behavior of graphs, yet possessing a certain level of BCU. Seeing in the context of Maple two holes at the points $x=\pm1$, one can use the Graphing Calculator to actually construct the holes through digital fabrication as shown in Figure 3.4 with the relations responsible for that fabrication in the input box of the program. As mentioned by the Conference Board of the

Mathematical Sciences (2012), in order to use computational tools strategically in teaching "to help students use them strategically in doing mathematics, teachers need to understand the mathematical aspects of these tools and their uses" (p. 2). Once such understanding is achieved, teachers can be expected "to provide a more engaging, student centered, and technology-enabled learning environment, and to promote greater diversity and creativity in learning" (Ministry of Education Singapore, 2012, p. 17).

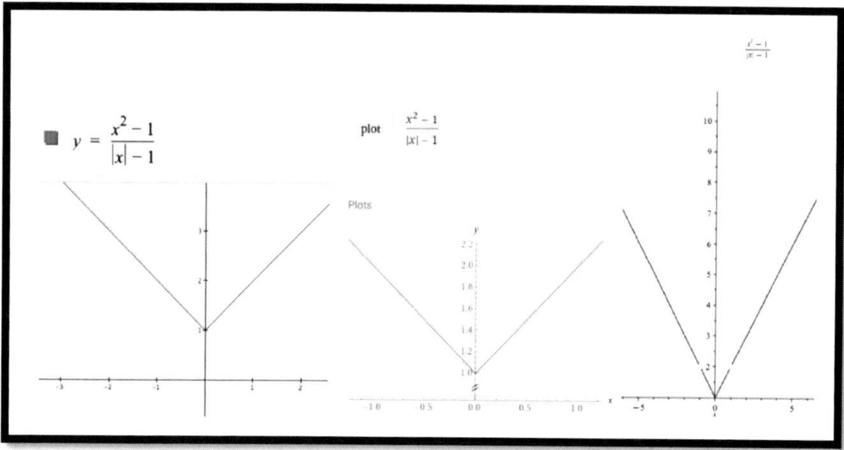

Figure 3.3. Observing technological inconsistency among Maple and two other tools.

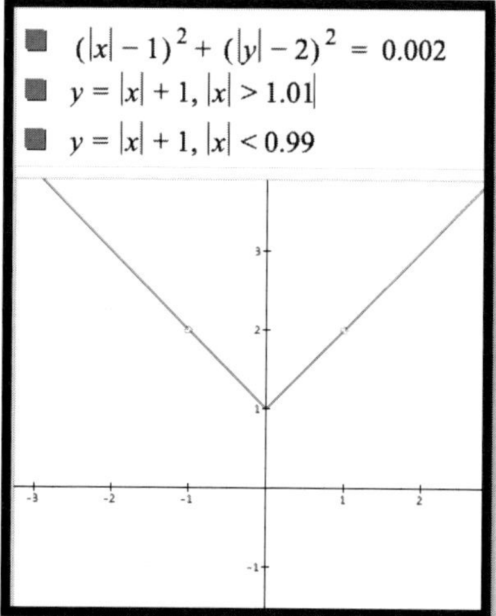

Figure 3.4. Computational triangulation fosters ACU.

At the same time, the above comparison of graphing by three different tools made in favor of Maple should not be taken to mean that one can always trust what Maple produces. When writing this book, the author was asked to review a paper submitted to an educational journal. In that paper, the use of Maple as a computational tool was discussed in the context of finding area bounded by the graphs of $y = x^2 + 1$ and $y = x + 3$. The image of the region presented in that paper looked like in Figure 3.5. It was clear that the parabola passing through the origin does not represent the graph of $y = x^2 + 1$ which must have the vertex at the point (0, 1) rather than at the origin. This kind of recognition required only BCU of parabolas and their graphs. When the author tried to use Maple to graph the two functions, surprisingly the result was the same. Different attempts to modify the code or just graph the parabola alone resulted in the same incorrect graph passing through the origin. This obvious glitch in the graphing (although $y = x^2 - 1$ was graphed by Maple correctly) was confirmed when the author graphed $y = x^2 + 1$ along with $y = 2x + 3$ (rather than with $y = x + 3$). The resulting graphs are shown in Figure 3.6 with the parabola $y = x^2 + 1$ correctly having the vertex at the point (0, 1). This example underscores the very idea of computational triangulation, be it graphing or any other type of digital computation. It also emphasizes the need for BCU of the behavior of basic functions (linear, quadratic, trigonometric, exponential), in order to decide which graph is correct when comparing possibly unlike results produced by different digital tools.

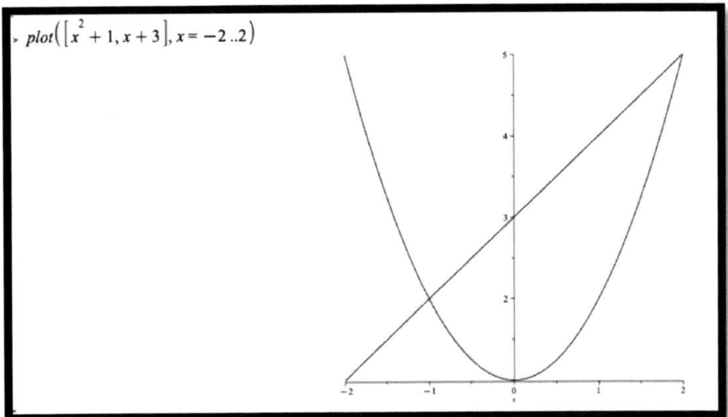

Figure 3.5. Maple yields incorrect graph of $y = x^2 + 1$ when graphed with $y = x + 3$.

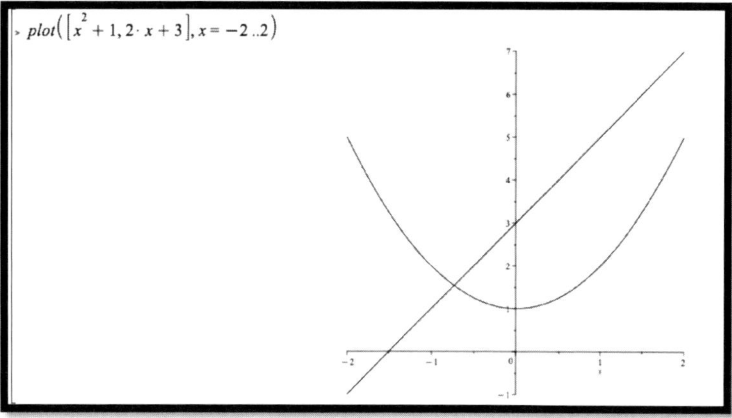

Figure 3.6. Maple yields correct graph of $y = x^2 + 1$ when graphed with $y = 2x + 3$.

3.5. About TITE Pedagogical Framework

Conceptual methodology introduced in this section and used in other chapters of the book follows the technology-immune/technology-enabled (TITE) framework (Abramovich, 2014). According to this framework, a TITE problem cannot be automatically solved by software (thus, it is *immune* from the direct use of technology), yet the role of software in solving the problem remains critical (thus, its solution is *enabled* by technology). An important aspect of the TITE problem-solving idea is that when the TI (technology-immune) part precedes the TE (technology- enabled) part of problem solving, depending on a problem, there are at least two outcomes available: an error in the TI part would either be missed or recognized through the action of the TE part. That is why, checking the result of a symbolic computation in a special case, the accuracy of which is semi-evident, should be included in a TITE problem solving (Abramovich, 2025a). Common Core State Standards (2010) emphasize the importance of investigating "special cases and simpler forms of the original problem in order to gain insight into its solution" (p. 6).

As an example, consider the situation when natural numbers are put in groups {1}, {2, 3}, {4, 5, 6}, {7, 8, 9, 10}, {11, 12, 13, 14, 15}, ... , the cardinalities of which increase by one, and the task is to find the sum of numbers in the n-th group. When solving this task as a TITE problem, one can notice that the largest number in each group can be expressed through the cardinalities of that group and the next one: e.g., $10 = \frac{1}{2} \times 4 \times 5$, and generalize that the largest number in the *n*-th group is equal to $\frac{1}{2} \times n \times (n+1) = t_n -$ the triangular number of rank *n*. In finding the smallest number in each group, one might erroneously assume that it differs from the largest number by the cardinality of the group and therefore, the smallest number in the *n*-th group is equal to $t_n - n$. Then, one can use Wolfram Alpha to find that

$$\sum_{i=\frac{1}{2}n(n+1)-n}^{i=\frac{1}{2}n(n+1)} i = \frac{1}{2}n^2(n+1)$$

is the sum of numbers in the *n*-th group. An error in the TI part (regarding the smallest number) can be missed unless one checks the result for a small value of *n*, say, *n* = 2. Whereas the TE result (based on the TI part) yields the number 6, the sum of numbers in the second group is the number 5. Of course, had one noticed that the smallest number in each group is one greater than the largest number in the previous group, the alternative summation

$$\sum_{i=\frac{1}{2}n(n-1)+1}^{i=\frac{1}{2}n(n+1)} i = \frac{1}{2}n(n^2+1)$$

would be confirmed in the same special case as $\frac{1}{2}2(2^2+1) = 5$. In both cases, attention to a special case has value from the TITE perspective. As an aside, note that the difference in the erroneous and correct sum of numbers in the *n*-th group is equal to the largest number in the previous group as $\frac{1}{2}n^2(n+1) - \frac{1}{2}n(n^2+1) = \frac{1}{2}n(n-1)$. Such custom of turning a mathematical error into a thinking device may be seen as an interplay between TI and TE parts of problem solving.

However, if (in a different task) one has to expand the binomial $(n+1)^3$ and mistakenly expands it as $n^3 + 3n^2 + 1$, the verification by Wolfram Alpha of the difference $(n+1)^3 - (n^3 + 3n^2 + 1)$ would not yield zero. Rather, the tool would yield 3*n*. This TE verification of the binomial expansion not only points at an error in the IT part, but it can be used to correct the error, as if $3n$ is subtracted from $(n+1)^3$ also, the result would be zero; that is, $(n+1)^3 = n^3 + 3n^2 + 3n + 1$ is the correct identity (binomial expansion).

Likewise, a TE activity may precede a TI one and while the latter would have no error, it may include erroneous interpretation of the results of the former. For example, using Wolfram Alpha as a TE summation activity, the following formula can be developed (see also Chapter 1, Section 1.3.3)

$$\sum_{i=1}^{n} i^5 = \frac{n^2(n+1)^2(2n^2+2n-1)}{12}.$$

This summation formula implies that its right-hand side, despite having a fractional form, represents an integer for any positive integer n. The special case $n = 2$ yields $\frac{2^2(2+1)^2(2\cdot 2^2+2\cdot 2-1)}{12} = 33 = 1^5 + 2^5$. One can further rewrite the fraction as the product of two fractions, $[\frac{n(n+1)}{2}]^2$ and $\frac{2n^2+2n-1}{3}$. Noticing that the first fraction is always an integer as the square of the n-th triangular number, one might assume that the second fraction is an integer as well for $n = 1, 2, 3, \ldots$. However, while this is true for $n = 1$, it is not true for $n = 2$. That is, whereas the TI part had no error in representing an integer expression as a product of two factors one of which is an integer, interpreting the second factor as an integer was incorrect. Once again, attention to special cases is a critical part of a mathematical practice. Nonetheless, the first factor may be a multiple of three, thereby providing needed division that was impossible for the second factor. Indeed, according to the last summation formula of the fifth powers of natural numbers, $1^5 + 2^5 = (\frac{2\times 3}{2})^2 \times \frac{8+4-1}{3} = 9 \times \frac{11}{3} = 33$. One can see same verification roles played by special cases when natural numbers were put in groups of increasing cardinality and when their fifth powers were added. In both problems, special cases provided a counterexample. More specifically, in the first case locating an error in the TI part was through the TE part; in the second case, an error in the TI interpretation was found by recourse to the foregoing TE part.

Another important aspect of the TITE problem-solving idea deals with its duality in a sense that whereas the TE part may inform the TI part, the latter, without ACU of the former, may lead desired generalization astray. For example, one of the author's students, when working on an activity intended to introduce Fibonacci numbers, correctly found that the integers 2, 3, 5, and 8 represent, respectively, the number of ways that one, two, three, and four two-colour (red/yellow) counters can be lined up so that no two red counters appear back-to-back. The student then entered the four integers into the input box of Wolfram Alpha, thus outsourcing to the tool both pattern recognition and inductive continuation of the sequence. The tool, proceeding from limited numeric information, instead of Fibonacci numbers, generated a sequence each term of which, beginning from the second term, augments triangular numbers 1, 3, 6 by two (3 = 1 + 2, 5 = 3 + 2, 8 = 6 + 2), continued the sequence with the integers 12 (= 10 + 2), 17 (= 15 + 2), 23 (= 21 + 2) and provided the closed-form sequence $a_n = \frac{n^2-n+4}{2}$. Indeed, $a_n = \frac{(n-1)n}{2} + 2 = t_{n-1} + 2$. This is not to blame the student for the lack of conceptual understanding of the activity with two-colour counters (that can be used to introduce Fibonacci numbers). Rather, this it to show how an attempt to generalize from limited numeric evidence while using single digital tool may be described algebraically in more than one correct way. This is where the TITE framework, ACU, and computational triangulation meet. One can check to see that entering the four numbers – 2, 3, 5, 8 – into the input box of OEIS® yields Fibonacci numbers as the only generalization. At the same time, simple googling of the four numbers offers both choices – Fibonacci numbers and triangular numbers increased by two. This also suggests (see Remark 7.3, Chapter 7, Section 7.4) that one needs to possess ACU of mathematics involved to recognize when having only four numbers may not be enough for digital generalization.

At the same time, a TI part can be used to improve the efficiency of a TE part, which, in turn, supports the advancement of a TI part at the level of generalization. For example, when using Wolfram Alpha to solve the equation $x + y + z = 10$ in positive integers that serve as the side lengths of a triangle with perimeter 10 linear units, the tool offers eight triples (satisfying the inequalities $x \geq y \geq z > 0$), only two of which, (4, 3, 3) and (4, 4, 2), satisfy the triangle inequality (see Chapter 1, Section 1.4, and Chapter 6, Section 6.4). However, when the inequality $x < y + z$ (enabling the largest side length to be smaller than the other two combined) is added to the code, it instructs Wolfram Alpha to generate only the above two triples, thereby improving computational efficiency of the TE part using the TI part and its advancement by integrating inequality-based reduction in computational problem solving.

3.6. Binet's Formula, Characteristic Equation, and Historical Connections

When using triangulation in solving mathematical problems, multiple approaches can be used to develop a single formula. Each approach can be described through the lens of the TITE framework. In Chapter 4, the following closed-form formula

$$F_n = \frac{1}{\sqrt{5}}[(\frac{1+\sqrt{5}}{2})^n - (\frac{1-\sqrt{5}}{2})^n], n = 1, 2, 3, ... \tag{3.4}$$

for Fibonacci numbers 1, 1, 2, 3, 5, 8, 13, ... will be used. Formula (3.4) is known as Binet's formula named after a French mathematician Jacques Philippe Marie Binet (1786-1856), although the same result was already known a century earlier to Abraham de Moivre (1667-1754), a French-born English mathematician. Whereas formula (3.4) can be obtained in a purely computational way, as it will be shown (through different notation) in Chapter 4, Section 4.2, using Maple by solving a recursive equation

$$F_{n+1} = F_n + F_{n-1}, F_1 = F_2 = 1, \tag{3.5}$$

it can also be developed by using the characteristic equation approach enhanced by the Wolfram Alpha symbolic computations. In other words, Binet's formula can be developed using the TITE framework introduced in the previous section. To this end, one can first solve the quadratic equation

$$x^2 - x - 1 = 0, \tag{3.6}$$

in which x^2, x and 1 replace, respectively, $F_{n+1}, F_n,$ and F_{n-1} in (3.5). The roots of quadratic equation (3.6) are $x_1 = \frac{1+\sqrt{5}}{2}$ and $x_2 = \frac{1-\sqrt{5}}{2}$. The correctness of the roots can be verified in a TI way by using formulas named after another French mathematician Franciscus Vieta (1540-1603) that connect the roots of a polynomial equation with its coefficients; that is, in the specific case of equation (3.6), $x_1 + x_2 = 1, x_1 \times x_2 = -1$. Indeed, $\frac{1+\sqrt{5}}{2} + \frac{1-\sqrt{5}}{2} = 1$ (the negation of the coefficient in x) and $\frac{1+\sqrt{5}}{2} \times \frac{1-\sqrt{5}}{2} = -1$ (the equation's free term). By setting $F_n = x(\frac{1+\sqrt{5}}{2})^n + y(\frac{1-\sqrt{5}}{2})^n$, one can set the system of equations (due to initial conditions set in recursive equation (3.5))

$$x(\frac{1+\sqrt{5}}{2}) + y(\frac{1-\sqrt{5}}{2}) = 1, x(\frac{1+\sqrt{5}}{2})^2 + y(\frac{1-\sqrt{5}}{2})^2 = 1,$$

and, moving to a TE part of the development of Binet's formula, solve it using Wolfram Alpha (Figure 3.7) to get $x = \frac{1}{\sqrt{5}}$ and $y = -\frac{1}{\sqrt{5}}$. Therefore, $F_n = \frac{1}{\sqrt{5}}[(\frac{1+\sqrt{5}}{2})^n - (\frac{1-\sqrt{5}}{2})^n]$ and this confirms Binet's formula (3.4) to be used in the following chapters of the book. Finally, to confirm the accuracy of the results of this section based on the notion of methodological triangulation (Denzin, 1970) as using more than one method in the study of the same thing, Binet's formula can be generated by Wolfram Alpha by entering into its input box the command "Binet's formula."

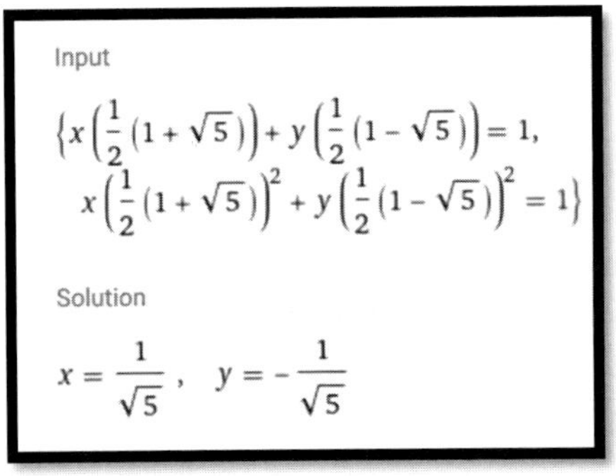

Input

$$\left\{x\left(\frac{1}{2}(1+\sqrt{5})\right)+y\left(\frac{1}{2}(1-\sqrt{5})\right)=1,\ x\left(\frac{1}{2}(1+\sqrt{5})\right)^2+y\left(\frac{1}{2}(1-\sqrt{5})\right)^2=1\right\}$$

Solution

$$x = \frac{1}{\sqrt{5}}, \quad y = -\frac{1}{\sqrt{5}}$$

Figure 3.7. Using *Wolfram Alpha* in solving a system of two linear equations.

Conclusion

The chapter reviewed the use of different digital tools regarding the accuracy of symbolic computations and graphic constructions they perform. It was pointed out that triangulating such computations and constructions is the modern-day necessity, especially in the context of mathematics teacher education with its focus on learning to teach the subject matter with multiple tools. The concepts of basic and advanced conceptual understanding were discussed through the lens of what sociologists call theoretical triangulation.

This discussion, illustrated by the notions of Einstellung effect and doubled experience, was supported by solving trigonometric and irrational equations. Practical and theoretical aspects of the TITE framework were discussed using examples from number theory followed by the development of Binet's formula. The scope of this chapter is appropriate to be included in problem-solving courses for prospective teachers of multiple grades. The next chapter will consider three scenarios of using the context of placing cookies on plates according to different recursive rules.

Chapter 4

Computational Triangulation Using Primary School Context

4.1. Introduction

In this chapter, computational triangulation is motivated by the primary school context dealing with placing cookies on plates using simple recursive rules based on the operations of addition and multiplication. One such rule is defined as follows: every number beginning from the third one is the sum of the previous two numbers allowing for the variation of the first two numbers. When the first two numbers are equal to one, we have the sequence of Fibonacci numbers 1, 1, 2, 3, 5, 8, 13, … . In terms of cookies on plates, placing a single cookie on each of the first two plates yields the number of cookies on each plate being a Fibonacci number. Allowing more than one cookie on the first (and/or second) plate, the rule of placing cookies on plates follows the development of the so-called Fibonacci-like numbers, a slight generalization of Fibonacci numbers. Leonardo Fibonacci (1170-1250), the most prominent Italian mathematician of his time, is credited, among other things, with the introduction of Hindu-Arabic number system into the Western world. Fibonacci numbers were also referred to as the series of Lamé, e.g., "la célèbre série de Lamé ou de *Fibonacci*" (Catalan, 1884, p. 8, italics in the original) named after a French mathematician and engineer Gabriel Lamé (1795–1870). The difference between Fibonacci and Fibonacci-like numbers is due to the first two terms – the latter numbers have at least the first term greater than one. One such example are Lucas numbers 2, 1, 3, 4, 7, 11, 18, 29,… , named after a French mathematician Edouard Lucas (1842–1891) who, according to Koshy (2001, p. 5) gave the famous sequence its modern name, Fibonacci numbers.

Other rules of placing cookies on plates will involve the joint use of addition and multiplication, operations with which young learners of mathematics are not only familiar but are assumed to be proficient in applications as they are expected to "understand concepts of area and relate area to multiplication and to addition" (Common Core State Standards, 2010,

p. 22). Just as Fibonacci (and Fibonacci-like) numbers demonstrate how integers can be represented as a combination of exponents with bases being irrational roots of quadratic equations (see Chapter 3, Section 3.6), similar exponential representations can be observed in the case of integers defined by other simple recursive definitions. Consequently, exploring the behavior of such integer sequences in the traditional (paper-and-pencil) way is computationally demanding and the use of multiple instruments under the umbrella of computational triangulation which provides rigor through confirming the compatibility of multiple representations is justifiable. Prospective teachers of mathematics should have experience with mathematical situations when simple repetitive operations result in sophisticated representational outcomes, like delineation of Fibonacci numbers through Binet's formula (Chapter 3, Section 3.6). This is consistent with the Australian perspective that whereas "challenging problems can be posed using basic content … effective use of digital technologies can enhance the relevance of the content and process for learning" (National Curriculum Board, 2008, p.14).

In what follows, it will be shown how such simple arithmetical rules can be described by linear recurrence equations that generate different integer sequences, both known and unknown to non-mathematicians, prospective teachers of mathematics included. More specifically, three arithmetic-based scenarios of placing cookies on plates will be considered through the combined lens of computational triangulation and the TITE methodology (Chapter 3, Section 3.5). In the first scenario, as was mentioned above, each plate beginning from the third one will have as many cookies as the previous two plates combined, the case of Fibonacci-like numbers. In the second scenario, each plate beginning from the third one will have twice as many cookies as the first plate of the pair of the previous plates plus the number of cookies on the second plate of the pair. This scenario will be connected to Jacobsthal numbers, named after a German mathematician Ernst Erich Jacobsthal (1882-1965). In the third scenario, each plate beginning from the third one will have twice as many cookies as the second plate of the pair of the previous plates plus the number of cookies on the first plate of the pair. This scenario will be connected to Pell numbers, named after an English mathematician John Pell (1611-1685). That is, each scenario can be enriched by a piece of history of the subject matter.

Through those contextually simple situations, different sequences will emerge allowing one to see how mathematical generalization brings about conceptually complicated outcomes hidden behind hands-on activities of placing cookies on plates. By recognizing what in mathematics and elsewhere is called "seeing the general in the particular" (e.g., Mason & Pimm, 1984), one can make connections among mathematical concepts, an educational emphasis of the modern-day learning and teaching standards in the United States (Common Core State Standards, 2010; Association of Mathematics Teacher Educators, 2017; Conference Board of the Mathematical Sciences, 2012), and elsewhere in North America (Ontario Ministry of Education, 2020; Western and Northern Canadian Protocol, 2008), South America (Felmer et al., 2014), Asia (Ministry of Education Singapore, 2020; Park & Choi-Koh, 2013), Africa (Department of Basic Education, 2018), Australia (National Curriculum Board, 2008), and Europe (Department for Education, 2013/2021).

4.2. Cookies on Plates: The First Scenario

Consider the case when several plates are lined up and the number of cookies on each plate beginning from the third one is the sum of the number of cookies on the previous two plates. This scenario is grounded in Fibonacci-like numbers. If x and y represent the number of cookies on the first and the second plates, respectively, then algebraic development of the number of cookies on the lined-up plates can be described, as a TI action, as follows:

$$x, y, x+y, x+2y, 2x+3y, 3x+5y, 5x+8y, 8x+13y, 13x+21y, \dots . \quad (4.1)$$

In sequence (4.1), the coefficients in x are the numbers 1, 1, 1, 2, 3, 5, 8, 13, … , which, beginning from the second term are consecutive Fibonacci numbers. Likewise, the coefficients in y are consecutive Fibonacci numbers 1, 1, 2, 3, 5, 8, 13, 21, … . That is, whereas sequence (4.1) generates different Fibonacci-like numbers by varying x and y, the coefficients in the linear combinations of x and y are Fibonacci numbers. This is an interesting property of Fibonacci-like numbers worthy of being singled out: any Fibonacci-like number beginning from the third can be represented as a linear combination of the first and the second ones with two consecutive Fibonacci numbers as coefficients. For example, the 7th and the 8th Lucas numbers are $18 = 5 \times 2 + 8 \times 1$ and $29 = 8 \times 2 + 13 \times 1$, where the initial values 2 and 1 form linear combinations with 5 and 8 as well as 8 and 13 being pairs of consecutive

Fibonacci numbers. At the same time, the sum of products $5 \times 2 + 13 \times 1 = 23$ and the number 23 is absent from the sequence of Lucas numbers because 5 and 13 are not consecutive Fibonacci numbers. Likewise, the sum of products $5 \times 2 + 3 \times 1 = 13$ and the number 13 is not a Lucas number because numbers 5 and 3, unlike 3 and 5, are consecutive Fibonacci number in the backward order.

The second step of the first scenario is to describe the Fibonacci number sequence recursively through the following difference equation

$$a_{n+1} = a_n + a_{n-1}, \; a_1 = a_2 = 1, \tag{4.2}$$

and then, using symbolic computations of Maple (a TE action), to solve this equation by finding its solution, $a_n = \frac{1}{\sqrt{5}}[(\frac{1+\sqrt{5}}{2})^n - (\frac{1-\sqrt{5}}{2})^n]$, in a closed form (which, as it was mentioned in Chapter 3, Section 3.6, bears the name of Binet). Such use of Maple is shown in Figure 4.1. The third step is to use Binet's formula in finding the number of cookies on the *n*-th plate. This step is carried out computationally using Maple (Figure 4.2) by setting $G_n = a_{n-2}x + a_{n-1}y$ to show that $G_5 = 2x + 3y$ – exactly the symbolic form of the number of cookies on the 5th plate as shown in sequence (4.1). That is, in general, the number of cookies a_n on the *n*-th plate is equal to

$$a_{n-2}x + a_{n-1}y = \frac{1}{\sqrt{5}}[(\frac{1+\sqrt{5}}{2})^{n-2} - (\frac{1-\sqrt{5}}{2})^{n-2}]x + \frac{1}{\sqrt{5}}[(\frac{1+\sqrt{5}}{2})^{n-1} - (\frac{1-\sqrt{5}}{2})^{n-1}]y, \tag{4.3}$$

where *x* and *y* are the number of cookies on the first and the second plates, respectively.

```
> rsolve({a(n + 1) = a(n) + a(n - 1), a(1) = 1, a(2) = 1}, {a})
```

$$\left\{ a(n) = \frac{\sqrt{5}\left(\frac{1}{2} + \frac{\sqrt{5}}{2}\right)^n}{5} - \frac{\sqrt{5}\left(\frac{1}{2} - \frac{\sqrt{5}}{2}\right)^n}{5} \right\}$$

Figure 4.1. Solving recursive equation (4.2) using Maple.

Remark 4.1. The use of Maple enabled solving recursive equation (4.2) and connecting its solution (Figure 4.1) to sequence (4.1) through computing (Figure 4.2) the right-hand side of relation (4.3) for $n = 5$, thereby demonstrating how a combination of TI and TE parts of the TITE methodology works. To continue explorations in a TI way, one can check to see that when the 5th plate has 13 cookies (setting $n = 5$ in formula (4.3) and computing $G(5)$ by Maple in Figure 4.2), the equation $2x + 3y = 13$ has two positive integer solutions: $x = 2$, $y = 3$, and $x = 5$, $y = 1$. Indeed, $2 \cdot 2 + 3 \cdot 3 = 13$ and $2 \cdot 5 + 3 \cdot 1 = 13$. Likewise, when the 5th plate has 15 cookies, the equation $2x + 3y = 15$ has two positive integer solutions: $x = 3$, $y = 3$, and $x = 6$, $y = 1$. Indeed, because in the last equation $2x$ must be divisible by 3 just as $3y$ and 15 are, the options for x are $x = 3$ implying $y = 3$, and $x = 6$ implying $y = 1$. (This TI strategy was not used in the case of 13 cookies because 13 is not divisible by either 2 or 3). Finally, when the 5th plate has 17 cookies, the equation $2x + 3y = 17$ has two positive integer solutions: $x = 4$, $y = 3$, and $x = 7$, $y = 1$. Indeed, $2 \cdot 4 + 3 \cdot 3 = 17$ and $2 \cdot 7 + 3 \cdot 1 = 17$. In particular, the three cases may serve as an illustration of what it might mean that "some equations … describe how two quantities vary together" (Conference Board of the Mathematical Sciences, 2012, p. 43).

> $a(n) := \frac{\sqrt{5}\left(\frac{1}{2}+\frac{\sqrt{5}}{2}\right)^n}{5} - \frac{\sqrt{5}\left(\frac{1}{2}-\frac{\sqrt{5}}{2}\right)^n}{5}$

$$a := n \mapsto \frac{\sqrt{5}\cdot\left(\frac{1}{2}+\frac{\sqrt{5}}{2}\right)^n}{5} - \frac{\sqrt{5}\cdot\left(\frac{1}{2}-\frac{\sqrt{5}}{2}\right)^n}{5}$$

> $G(n) := a(n-2)\cdot x + a(n-1)\cdot y$

$$G := n \mapsto a(n-2)\cdot x + a(n-1)\cdot y$$

> $G(5)$

$$\left(\frac{\sqrt{5}\left(\frac{1}{2}+\frac{\sqrt{5}}{2}\right)^3}{5} - \frac{\sqrt{5}\left(\frac{1}{2}-\frac{\sqrt{5}}{2}\right)^3}{5}\right)x + \left(\frac{\sqrt{5}\left(\frac{1}{2}+\frac{\sqrt{5}}{2}\right)^4}{5} - \frac{\sqrt{5}\left(\frac{1}{2}-\frac{\sqrt{5}}{2}\right)^4}{5}\right)y$$

> simplify(%)

$$2x + 3y$$

Figure 4.2. Maple confirms $2x + 3y$ cookies on the 5th plate using solution of (4.2).

Furthermore, one can observe that in each of the three cases (13, 15, and 17 cookies on the 5th plate, with two solutions in each case), adding cookies from the first two plates in the first solution yields the cookies on the 1st plate of the second solution with a single cookie on the second plate. Indeed, in the case of 13 cookies we have $2 + 3 = 5$, in the case of 15 cookies we have $3 + 3 = 6$, and in the case of 17 cookies we have $4 + 3 = 7$, and a single cookie can aways be observed on the 2nd plate of the second solution in all three cases.

This observation can be explained by noting that when x and y are the number of cookies on the first two plates, respectively, and $x + y$ and 1 represent another arrangement of cookies on the first two plates, then the development of cookies in the latter case would be as follows: $x + y$, 1, $x + y + 1$, $x + y + 2$, $2x + 2y + 3$. That is, $2x + 2y + 3 = 2x + 3y$ whence $y = 3$ – the number of cookies on the 2nd plate of the first solution in each of the three cases. One can see that whenever, in the first solution, the number of cookies on the 2nd plate is 3, the number of cookies on the 1st plate of the second solution is equal to as many cookies as we have on the first two plates of the first solution and a single cookie on the second plate. One can also check to see that in each of the three cases having 28, 31 and 34 cookies on the 6th plate (with exactly two solutions in each case), whenever in the first solution the number if cookies on the second plate is 5, the number of cookies on the first plate of the second solution is the sum of the number of cookies on the first two plates of the first solution and 2 cookies on the second plate. Indeed, the first two plates in the cases of 28, 31, and 34 cookies on the 6th plate are {(1, 5), (6, 2)}, {(2, 5), (7, 2)}, and {(3, 5), (8, 2)}, respectively.

Remark 4.2. Applying what sociologists call "triangulation *between* methods" (McFee, 1992, p. 217; italics in the original), the correctness of formula (4.3) can be verified in a TE way (i.e., computationally) by using the Graphing Calculator. As mentioned in Remark 4.1, to have the 5th plate with 15 cookies, the first two plates must have either 3 and 3 cookies or 6 and 1 cookies. In the coordinate plane (x, y) one can construct the points (3, 3) and (6, 1) to see, according to formula (4.3), that they belong to the straight line defined by the equation

$$\frac{1}{\sqrt{5}}[(\frac{1+\sqrt{5}}{2})^{n-2} - (\frac{1-\sqrt{5}}{2})^{n-2}]x + \frac{1}{\sqrt{5}}[(\frac{1+\sqrt{5}}{2})^{n-1} - (\frac{1-\sqrt{5}}{2})^{n-1}]y = 15 \quad (4.4)$$

when $n = 5$. This is confirmed by the graph of Figure 4.3 (left) where, in addition, the constructed points represent the graphs of the inequalities $(x - 3)^2 + (y - 3)^2 < 0.01$ and $(x - 6)^2 + (y - 1)^2 < 0.01$ which define small disks with radii 0.1 centered at the above two points (see Chapter 3, Figure 3.4). Alternatively, using "triangulation … *within* a method" (McFee, 1992, p. 217; italics in the original), one can graph the equation $\frac{y-3}{1-3} = \frac{x-3}{6-3}$ of the straight line passing through the two points (Figure 4.3, right; hereafter, y' and x' represent notation in the Graphing Calculator, not derivatives), another way of confirming that equation (4.4) turns into $2x + 3y = 15$ when $n = 5$. Likewise, to have the 6th plate with 28 cookies, the first two plates must have

1 and 5 cookies or 6 and 2 cookies. Figure 4.4 (left) shows the graph of equation (4.4) in which $n = 6$ and the right-hand side is 28.

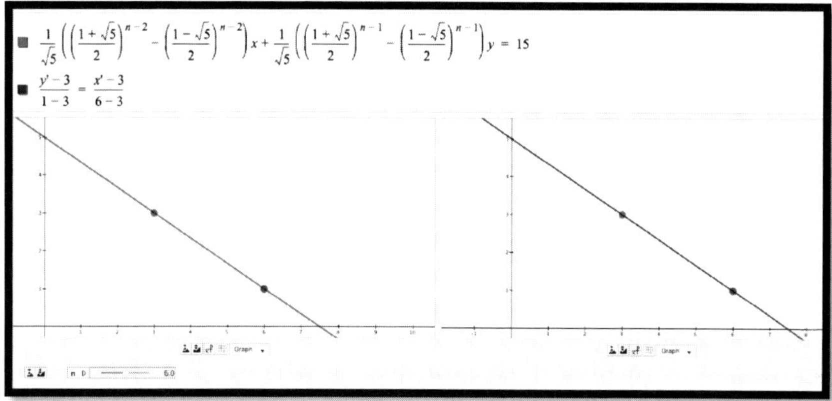

Figure 4.3. Two ways of demonstrating where the points (3, 3) and (6, 1) reside.

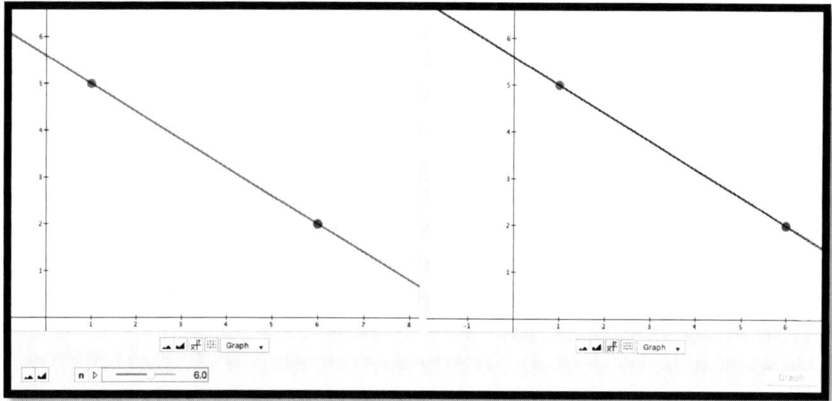

Figure 4.4. Two ways of demonstrating where the points (1, 5) and (6, 2) reside.

Figure 4.4 (right) shows the graph of the equation $\frac{y-5}{2-5} = \frac{x-1}{6-1}$ (which is equivalent to $3x + 5y = 28$). In both cases, the points (1, 5) and (6, 2) are digitally fabricated. These constructions can also be seen as triangulation by data sources (Denzin, 1970) when the same method of using the Graphing Calculator is involved, but mathematically data sources are different. The first data source is Binet's formula used to represent Fibonacci numbers as coefficients of linear equations generating Fibonacci-like numbers. The second data source is the equation of the straight line passing through two points in the plane. As two points, defining a straight line in the plane, belong to the same line, one uses graphing to "identify when two expressions are equivalent" (Common Core State Standards, 2010, p. 44). One can see that the method of graphing can be employed by the learners of mathematics "to maximum theoretical advantage" (Denzin, 1970, p. 301). Put another way, the method of graphing represents a combination of TI analysis (constructing equations and inequalities using mathematics) and TE exploration (graphing equations and inequalities using technology) and maximizes the use of formal knowledge in support of digital fabrication.

4.3. Cookies on Plates: The Second Scenario

Consider the case when the number of cookies on each plate beginning from the third one has twice as many cookies as the first plate of the pair of the previous plates plus the number of cookies on the second plate of the pair. If x and y represent the number of cookies on the first and the second plates, respectively, the algebraic development of the number of cookies on the lined-up plates can be described, as a TI action, through the sequence

$$x, y, 2x + y, 2x + 3y, 6x + 5y, 10x + 11y, 22x + 21y, \ldots. \qquad (4.5)$$

Coefficients in x in sequence (4.5) can be described through the sequence a_n as follows:

$a_1 = 1,\ a_2 = 0, a_3 = 2,\ a_4 = 2, a_5 = 6,\ a_6 = 10, a_7 = 22,\ a_8 = 42, \ldots.$

Recursively, we have $a_1 = 1,\ a_2 = 2\,a_1 - 2,\ a_3 = 2\,a_2 + 2, a_4 = 2\,a_3 - 2,\ a_5 = 2\,a_4 + 2, a_6 = 2\,a_5 - 2, a_7 = 2\,a_6 + 2,\ a_8 = 2\,a_7 - 2$, and so on. In general,

$$a_n = 2a_{n-1} + 2(-1)^{n+1}, a_1 = 1. \qquad (4.6)$$

As a TE action, one can use Wolfram Alpha (Figure 4.5) to find the closed-form formula for the sequence a_n by solving recursive equation (4.6).

The formula is $a_n = \frac{2^n - 4(-1)^n}{6}$. A similar, but symbolically different formula, $f_n = \frac{2^n}{6} - \frac{2(-1)^n}{3}$, can be generated by Maple (Figure 4.6). To show that both tools generate identical results, computational triangulation of the second order can be applied. A simple algebraic transformation is carried out by Wolfram Alpha in Figure 4.7 to show that $a_n - f_n = 0$.

Coefficients in *y* in sequence (4.5) can be described through the sequence b_n as follows:

$$b_1 = 0, b_2 = 1, b_3 = 1, b_4 = 3, b_5 = 5, b_6 = 11, b_7 = 21, b_8 = 43, \ldots .$$

In a TI fashion, one can note that these numbers are either one greater or one smaller than the corresponding coefficients in *x*. That is, $b_n = a_n + (-1)^n$. Therefore,

$$b_n = \frac{2^n - 4(-1)^n}{6} + (-1)^n = \frac{2^n - 4(-1)^n + 6(-1)^n}{6} = \frac{2^n + 2(-1)^n}{6}.$$

Finally, the number of cookies on the *n*-th plate in the second scenario can be described by the formula

$$f_n(x, y) = \frac{2^n - 4(-1)^n}{6} x + \frac{2^n + 2(-1)^n}{6} y. \tag{4.7}$$

Formula (4.7), which is a generalization of sequence (4.5), will be discussed in Chapter 5 in connection with allowing for non-integer values of *x*, *y*, and $f_n(x, y)$ in the context of cookies on plates.

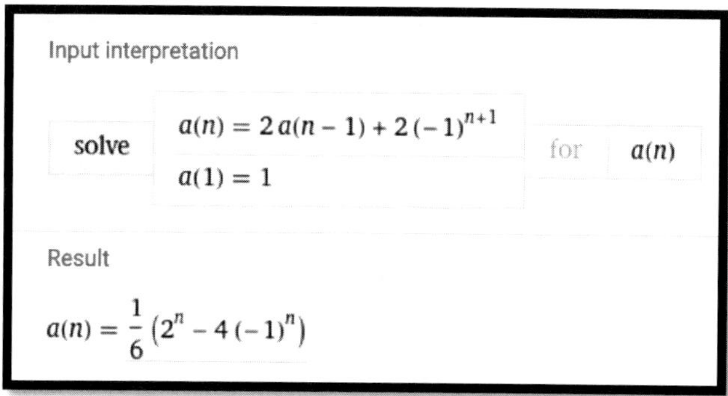

Figure 4.5. Solving recursive equation (4.6) using Wolfram Alpha.

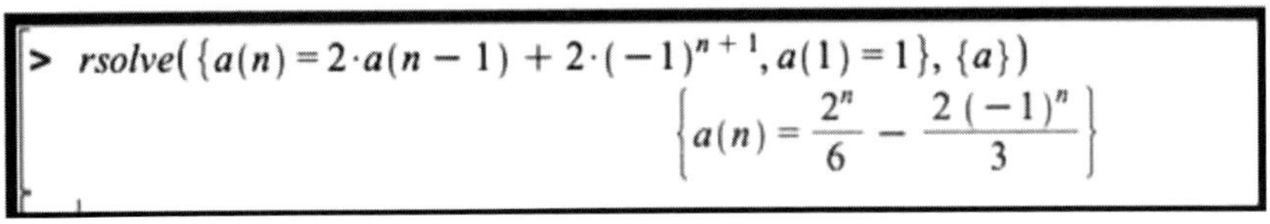

```
> rsolve({a(n) = 2·a(n − 1) + 2·(−1)^(n+1), a(1) = 1}, {a})
```

$$\left\{a(n) = \frac{2^n}{6} - \frac{2(-1)^n}{3}\right\}$$

Figure 4.6. Solving recursive equation (4.6) using Maple.

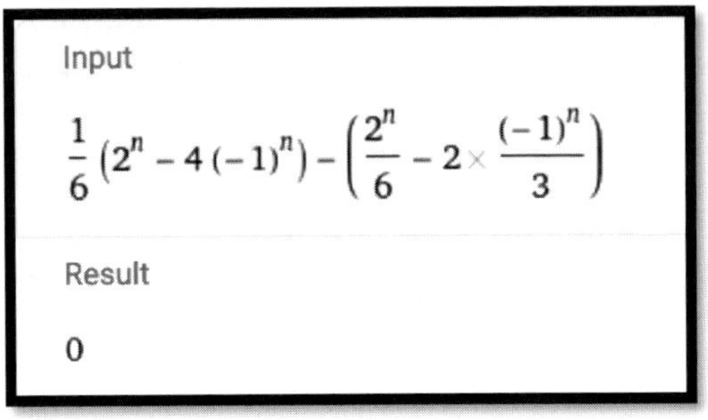

Input

$$\frac{1}{6}(2^n - 4(-1)^n) - \left(\frac{2^n}{6} - 2\times\frac{(-1)^n}{3}\right)$$

Result

0

Figure 4.7. Computational triangulation of the second order for the second scenario.

Remark 4.3. The numbers a_n (and, consequently, b_n) can be connected to Jacobsthal numbers $J_n = \frac{2^n-(-1)^n}{3} = \{1, 1, 3, 5, 11, 21, 43, \ldots\}$ as follows:

$$a_n = \frac{2^n - 4(-1)^n}{6} = \frac{2^n - (-1)^n - 3(-1)^n}{6} = \frac{2^n - (-1)^n}{6} - \frac{3(-1)^n}{6}$$
$$= \frac{1}{2}\left(\frac{2^n-(-1)^n}{3} - (-1)^n\right) = \frac{1}{2}(J_n - (-1)^n).$$

Therefore, $b_n = a_n + (-1)^n = \frac{1}{2}(J_n - (-1)^n) + (-1)^n = \frac{1}{2}(J_n + (-1)^n)$.

Alternatively, the number of cookies on the *n*-th plate can be represented through Jacobsthal numbers J_n as follows:

$$f_n(x, y) = \frac{1}{2}(J_n - (-1)^n)x + \frac{1}{2}(J_n + (-1)^n)y.$$

When $n = 7$, we have $J_7 = 43$ and, therefore,

$$f_7(x, y) = \frac{1}{2}(43 + 1)x + \frac{1}{2}(43 - 1)y = 22x + 21y.$$

That is, $f_7(x, y)$ is exactly the 7th term of sequence (4.5). Furthermore, $f_7(1,1) = 43 = J_7$ – the 7th Jacobsthal number, representing the number of cookies on the 7th plate in the context of the second scenario when the first two plates have a single cookie each.

Remark 4.4. Comparing the second scenario to the first scenario, one can see that 15 cookies on the 4th plate within the former scenario coincide with 15 cookies on the 5th plate within the latter scenario (see Remark 4.1) and this is the only case when algebraic representations of cookies on the two plates are the same. Indeed, when $n = 4$ (the 4th plate in the second scenario) we have $f_4(x, y) = \frac{16-4}{6}x + \frac{16+2}{6}y = 2x + 3y$, and when $n = 5$ (the 5th plate in the first scenario) we have (as shown by Maple in Figure 4.2)

$$a_5 = \frac{1}{\sqrt{5}}[(\frac{1+\sqrt{5}}{2})^{5-2} - (\frac{1-\sqrt{5}}{2})^{5-2}]x + \frac{1}{\sqrt{5}}[(\frac{1+\sqrt{5}}{2})^{5-1} - (\frac{1-\sqrt{5}}{2})^{5-1}]y = 2x + 3y.$$

Put another way, the numbers 2 and 3 are the only consecutive Fibonacci numbers greater than one that are also consecutive natural numbers.

Remark 4.5. Consider the equation $6x + 5y = 41$ which represents the case when in the second scenario the number of cookies on the 5th plate is 41 – the smallest number for which the equation has exactly two positive integer solutions. Solving this equation using Wolfram Alpha yields $x = 1, y = 7$ and $x = 6, y = 1$, where, as one can note, $7 = 6 + 1$. The number of cookies in each case develops, respectively, as follows: 1, 7, 9 $(= 2 \times 1 + 7)$, 23 $(= 2 \times 7 + 9)$, 41 $(= 2 \times 9 + 23)$ and 6, 1, 13 $(= 2 \times 6 + 1)$, 15 $(= 2 \times 1 + 13)$, 41 $(= 2 \times 13 + 15)$. Figure 4.8 (left; triangulation between methods) shows how the Graphing Calculator can graph the equation with the right-hand side of (4.7) equal to 41 in the case $n = 5$ (the 5th plate); the graph passes through the points (1, 7) and (6, 1). Furthermore, Figure 4.8 (right; triangulation within a method) shows that the graph of the equation $\frac{y-7}{1-7} = \frac{x-1}{6-1}$, an equivalent form of which is the equation $6x + 5y = 41$, also passes through the above two points. Alternatively, as mentioned in Remark 4.2, such use of the Graphing Calculator can be described as triangulation by data sources.

To explain the relation $7 = 6 + 1$ mentioned above, note that in the equation $6x + 5y = 41$ the pairs (1, 7) and (6, 1) make the first addend, $6x$, equal to 6 and 36 (increase of 30), and the second addend, $5y$, equal to 35 and

5 (decrease of 30), respectively, where 30 = LCM (6, 5). Also, $41 = 6 \times 6 + 5 \times 1 = (5 + 1) \times 6 + 5 \times 1 = 5 \times 6 + 5 \times 1 + 6 = 5 \times (6 + 1) + 6 = 6 \times 1 + 5 \times 7 = 41$. In terms of cookies on plates, if 41 cookies can be reached on the 5th plate by placing one and seven cookies on the 1st and the 2nd plates, respectively, then seven cookies on the 2nd plate can be split into six and one cookies to be placed on the 1st and the 2nd plates, respectively, to also reach 41 cookies on the 5th plate.

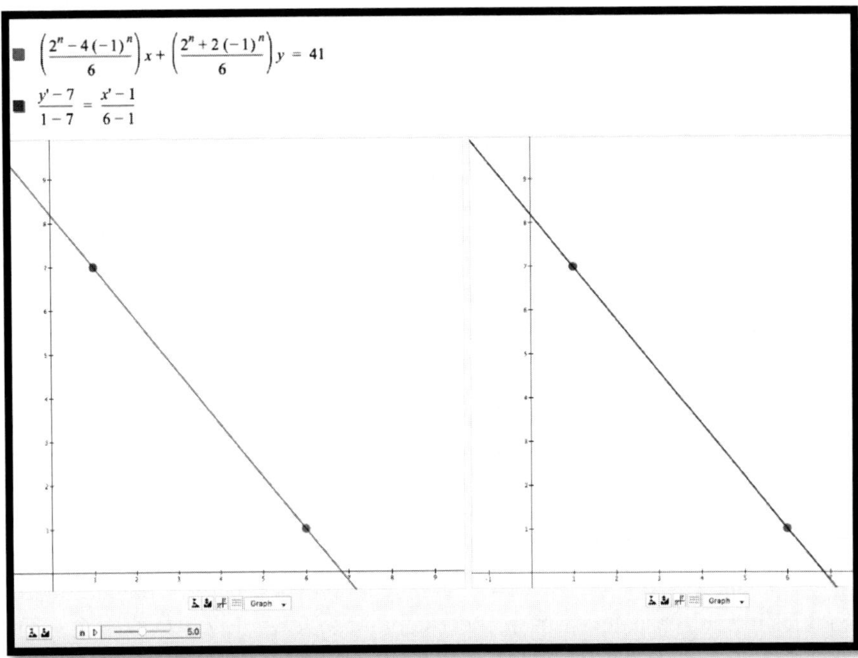

Figure 4.8. Two ways of demonstrating where the points (1, 7) and (6, 1) reside.

Likewise, using Wolfram Alpha, one can find out that the equation $6x + 5y = 46$ has exactly two positive integer solutions, (1, 8) and (6, 2), where, as one can note, 8 = 6 + 2. The number of cookies in each case develops, respectively, as follows: 1, 8, 10 $(= 2 \times 1 + 8)$, 26 $(= 2 \times 8 + 10)$, 46 $(= 2 \times 10 + 26)$ and 6, 2, 14 $(= 2 \times 6 + 2)$, 18 $(= 2 \times 2 + 14)$, 46 $(= 2 \times 14 + 18)$. To explain the relation 8 = 6 + 2, note that in the equation $6x + 5y = 46$ the pairs (1, 8) and (6, 2) make the first addend, $6x$, equal to 6 and 36 (increase of 30), and the second addend, $5y$, equal to 40 and 10 (decrease of 30), respectively, where 30 = LCM (6, 5). Also, $46 = 6 \times 6 + 5 \times 2 =$

$(5+1)\times 6+5\times 2=5\times 6+5\times 2+6=5\times(6+2)+6=6\times 1+5\times 8=46$. In terms of cookies on plates, if 46 cookies can be reached on the 5th plate by placing one and eight cookies on the 1st and the 2nd plates, respectively, then eight cookies on the 2nd plate can be split into six and two cookies to be placed on the 1st and the 2nd plates, respectively, to also reach 46 cookies on the 5th plate (of another set of plates).

Remark 4.6. The second scenario with a single cookie on each of the first two plates can also be described through the recursive equation of the second order

$$g_{n+1}=g_n+2g_{n-1}, g_1=g_2=1\,. \tag{4.8}$$

Solving equation (4.8) by Wolfram Alpha yields $g_n=c_1(-1)^n+c_2 2^n$. Setting $n=1$ and $n=2$ in the last formula and considering the initial conditions of (4.8) yields $-c_1+2c_2=1$ and $c_1+4c_2{=}1$. Adding the last two equations yields $6c_2=2$ or $c_2=\frac{1}{3}$ and $c_1=\frac{2}{3}-1=-\frac{1}{3}$ whence

$$g_n=\frac{(-1)^{n+1}+2^n}{3}\,. \tag{4.9}$$

Note that setting $x=y=1$ in formula (4.7) yields $f_n(1,1)=\frac{2^n-4(-1)^n}{6}+\frac{2^n+2(-1)^n}{6}=\frac{2^{n+1}-2(-1)^n}{6}=\frac{(-1)^{n+1}+2^n}{3}$ which is exactly the right-hand side of formula (4.9). That is, $g_n=f_n(1,1)$.

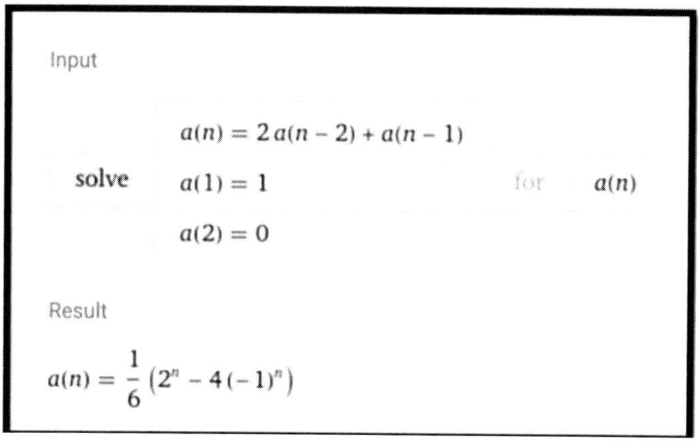

Input

solve $a(n)=2a(n-2)+a(n-1)$, $a(1)=1$, $a(2)=0$ for $a(n)$

Result

$$a(n)=\frac{1}{6}\left(2^n-4(-1)^n\right)$$

Figure 4.9. The second scenario with one and no cookies on the 1st and 2nd plates, respectively.

Furthermore, formula (4.9) is identical to the formula for Jacobsthal numbers mentioned in Remark 4.3. Finally, as shown in Figure 4.9 created by Wolfram Alpha, the formula $a_n = \frac{2^n - 4(-1)^n}{6}$ is a closed-form solution to the equation $a_n = a_{n-1} + 2a_{n-2}, a_1 = 1, a_2 = 0$, describing the second scenario when the 1st plate has a single cookie and the 2nd plate is empty.

4.4. Cookies on Plates: The Third Scenario

Consider the case when the number of cookies on each plate beginning from the third one has twice as many cookies as the second plate of the pair of the previous two plates plus the number of cookies on the first plate of the pair. If *x* and *y* represent the number of cookies on the first and the second plates, respectively, the algebraic development of the number of cookies on the lined-up plates can be described, as a TI action, through the sequence

$$x, y, x + 2y, 2x + 5y, 5x + 12y, 12x + 29y, 29x + 70y, 70x + 169y, \dots . \qquad (4.10)$$

Coefficients in *x* in (4.10) can be described through the sequence a_n as follows:

$$a_1 = 1,\ a_2 = 0, a_3 = 1,\ a_4 = 2, a_5 = 5,\ a_6 = 12, a_7 = 29,\ a_8 = 70, \dots .$$

Recursively,

$a_1 = 1,\ a_2 = 0, a_3 = 2\,a_2 + a_1, a_4 = 2\,a_3 + a_2,\ a_5 = 2\,a_4 + a_3,\ a_6 = 2\,a_5 + a_4$, and so on. In general, we have the following recursive equation, known as Pell recurrence (Koshy, 2014),

$$a_n = 2a_{n-1} + a_{n-2}, a_1 = 1, a_2 = 0. \qquad (4.11)$$

Coefficients in *y* in (4.10), known as Pell numbers (Koshy, 2014), can be described as follows:

$$b_1 = 0,\ b_2 = 1, b_3 = 2,\ b_4 = 5, b_5 = 12,\ b_6 = 29, b_7 = 70,\ b_8 = 169, \dots .$$

That is, $b_n = a_{n+1}$. As a TE action, using Maple (Figure 4.10) one can solve recursive equation (4.11) to get

$$a_n = (-1 - \tfrac{3\sqrt{2}}{4})(-\sqrt{2} + 1)^n + (-1 + \tfrac{3\sqrt{2}}{4})(\sqrt{2} + 1)^n.$$

Consequently,

$$b_n = (-1 - \tfrac{3\sqrt{2}}{4})(-\sqrt{2} + 1)^{n+1} + (-1 + \tfrac{3\sqrt{2}}{4})(\sqrt{2} + 1)^{n+1}.$$

The last formula, found through a combination of TI (observation and analysis) and TE (the use of Maple) activities, is used to generate the first 20 Pell numbers shown in Figure 4.11. Likewise, the closed-form formula for a_n can be used to generate same numbers by changing n from 2 to 21.

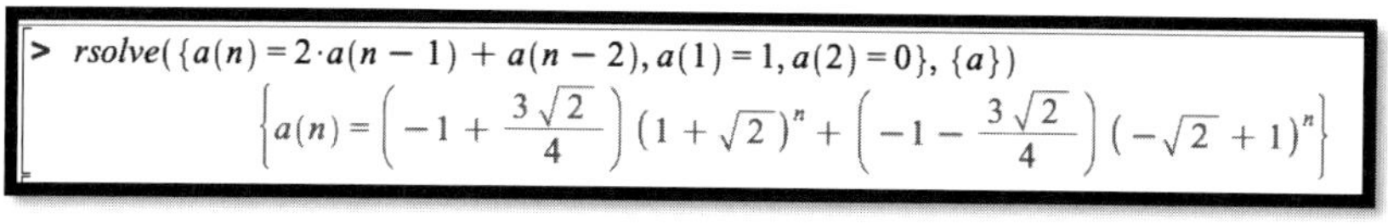

> $rsolve(\{a(n) = 2 \cdot a(n-1) + a(n-2), a(1) = 1, a(2) = 0\}, \{a\})$

$$\left\{a(n) = \left(-1 + \frac{3\sqrt{2}}{4}\right)(1 + \sqrt{2})^n + \left(-1 - \frac{3\sqrt{2}}{4}\right)(-\sqrt{2} + 1)^n\right\}$$

Figure 4.10. Solving recursive equation (4.11) using Maple.

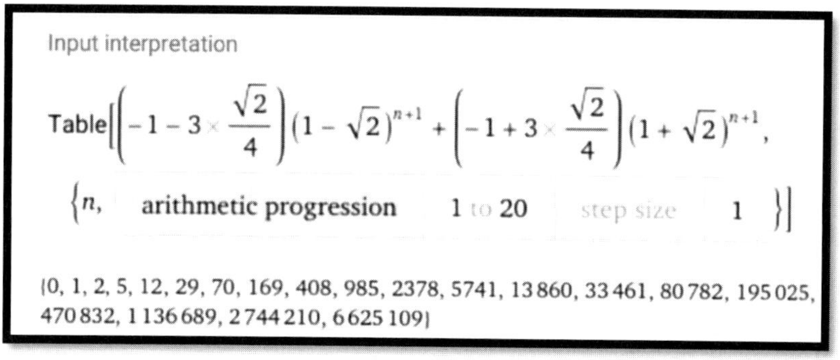

Input interpretation

$$\text{Table}\left[\left(-1 - 3 \times \frac{\sqrt{2}}{4}\right)(1 - \sqrt{2})^{n+1} + \left(-1 + 3 \times \frac{\sqrt{2}}{4}\right)(1 + \sqrt{2})^{n+1},\right.$$

$\{n,$ arithmetic progression 1 to 20 step size 1 $\}]$

{0, 1, 2, 5, 12, 29, 70, 169, 408, 985, 2378, 5741, 13860, 33461, 80782, 195025, 470832, 1136689, 2744210, 6625109}

Figure 4.11. Using Wolfram Alpha to generate the first 20 Pell numbers.

Using formulas for a_n and b_n, the number of cookies on the n-th plate can be described through a linear combination of x and y as follows:

$$a_n x + b_n y =$$
$$[(-1-\frac{3\sqrt{2}}{4})(-\sqrt{2}+1)^n + \left(-1+\frac{3\sqrt{2}}{4}\right)(\sqrt{2}+1)^n]x +$$
$$[(-1-\tfrac{3\sqrt{2}}{4})(-\sqrt{2}+1)^{n+1} + \left(-1+\tfrac{3\sqrt{2}}{4}\right)(\sqrt{2}+1)^{n+1}]y. \qquad (4.12)$$

Using Wolfram Alpha (Figure 4.12) in solving recursive equation (4.11) yields a different form of solution in comparison with Maple (Figure 4.10). Because symbolic forms of the solutions of recursive equation (4.11) provided by the two digital instruments look different, one can use one of them, Maple, to carry out computational triangulation of the second order. Such confirmation of identity of the two forms by simplifying their difference to zero is shown in Figure 4.13.

Furthermore, one can use Wolfram Alpha (Figure 4.14) to verify relation (4.12) in the case $n = 5$, expecting to get the fifth term of sequence (4.10). As mentioned in Chapter 3, Section 3.5, checking results of symbolic computations is an important part of the TITE problem-solving methodology that activities of this chapter follow. Indeed,

$$\begin{aligned} a_5 x + b_5 y = {} & [(-1-\frac{3\sqrt{2}}{4})(-\sqrt{2}+1)^5 + \left(-1+\frac{3\sqrt{2}}{4}\right)(\sqrt{2}+1)^5]x + [(-1 \\ & -\frac{3\sqrt{2}}{4})(-\sqrt{2}+1)^{5+1} + \left(-1+\frac{3\sqrt{2}}{4}\right)(\sqrt{2}+1)^{5+1}]y \\ = {} & 5x + 12y. \end{aligned}$$

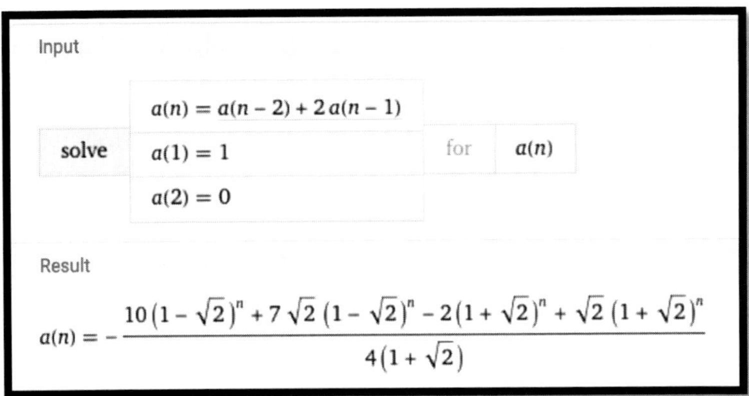
Input

solve $a(n) = a(n-2) + 2a(n-1)$, $a(1) = 1$, $a(2) = 0$ for $a(n)$

Result

$$a(n) = -\frac{10(1-\sqrt{2})^n + 7\sqrt{2}(1-\sqrt{2})^n - 2(1+\sqrt{2})^n + \sqrt{2}(1+\sqrt{2})^n}{4(1+\sqrt{2})}$$

Figure 4.12. Solving recursive equation (4.11) using Wolfram Alpha.

```
> rsolve({a(n) = 2·a(n − 1) + a(n − 2), a(1) = 1, a(2) = 0}, {a})
```

$$\left\{a(n)=\left(-1+\frac{3\sqrt{2}}{4}\right)(1+\sqrt{2})^n+\left(-1-\frac{3\sqrt{2}}{4}\right)(-\sqrt{2}+1)^n\right\}$$

```
> −(10·(1 − √2)^n + 7·√2·(1 − √2)^n − 2·(1 + √2)^n + √2·(1 + √2)^n)/(4·(1 + √2))
  −(−1 − 3√2/4)(−√2 + 1)^n − (−1 + 3√2/4)(1 + √2)^n
```

$$-\frac{10(-\sqrt{2}+1)^n+7(-\sqrt{2}+1)^n\sqrt{2}-2(1+\sqrt{2})^n+(1+\sqrt{2})^n\sqrt{2}}{4+4\sqrt{2}}-\left(-1-\frac{3\sqrt{2}}{4}\right)(-\sqrt{2}+1)^n-\left(-1+\frac{3\sqrt{2}}{4}\right)(1+\sqrt{2})^n$$

```
> simplify(%)
```

$$0$$

Figure 4.13. Computational triangulation between Wolfram Alpha and Maple using the latter.

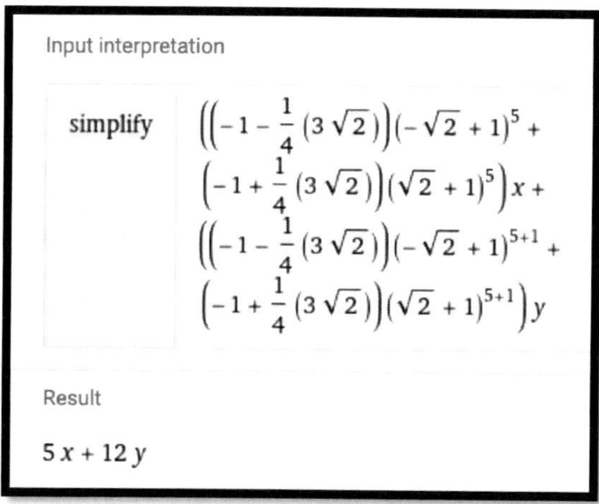

Figure 4.14. Verifying the general result in a specific case.

Likewise, one can use the Graphing Calculator (Figure 4.15, left) to show that the relation

$$[(-1-\frac{3\sqrt{2}}{4})(-\sqrt{2}+1)^n+\left(-1+\frac{3\sqrt{2}}{4}\right)(\sqrt{2}+1)^n]x+$$
$$[(-1-\frac{3\sqrt{2}}{4})(-\sqrt{2}+1)^{n+1}+\left(-1+\frac{3\sqrt{2}}{4}\right)(\sqrt{2}+1)^{n+1}]y=77$$

is the equation of the straight line that passes though the points (1, 6) and (13, 1) when $n = 5$. The two points represent integer solutions of the equation $5x +$

$12y = 77$ as shown in Figure 4.16 generated by Wolfram Alpha. Note that 77 is the smallest right-hand side of the last equation providing exactly two positive integer solutions.

As a triangulation within a graphing method, one can graph the equation $\frac{y-6}{1-6} = \frac{x-1}{13-1}$ of the straight line that passes through the two points (Figure 4.15, right), an equivalent form of which is the equation $5x + 12y = 77$.

$$a_5x + b_5y = [(-1 - \frac{3\sqrt{2}}{4})(-\sqrt{2} + 1)^5 + (-1 + \frac{3\sqrt{2}}{4})(\sqrt{2} + 1)^5]x + [(-1 - \frac{3\sqrt{2}}{4})(-\sqrt{2} + 1)^{5+1} + (-1 + \frac{3\sqrt{2}}{4})(\sqrt{2} + 1)^{5+1}]y = 77.$$

Once again, the graphing method can be used by "triangulating data sources [solution of a recurrence equation and the equation of the straight line passing through two points] … to maximum theoretical advantage" (Denzin, 1970, p. 301) as a teaching method in the digital era. In other words, the combination of TI and TE explorations when theory is used to motivate and support computations and computations enable verification and advancement of theory, maximizes teachers' learning of mathematical ideas in the digital era.

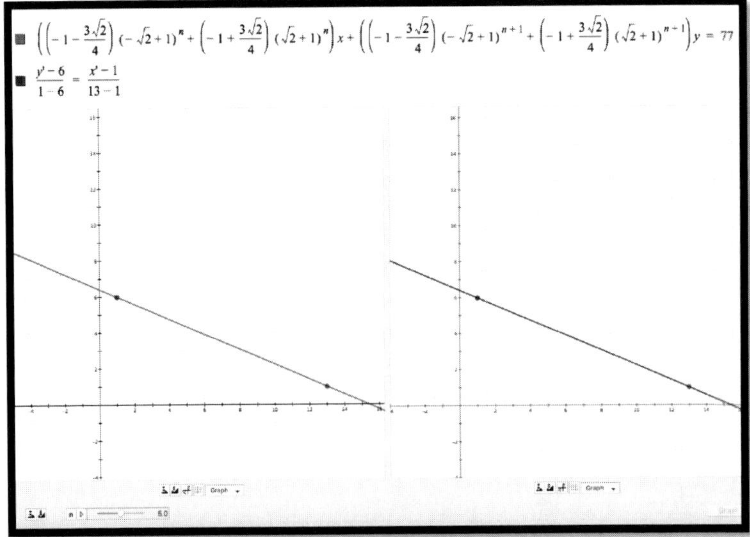

Figure 4.15. Two ways of demonstrating where the points (1, 6) and (13, 1) reside.

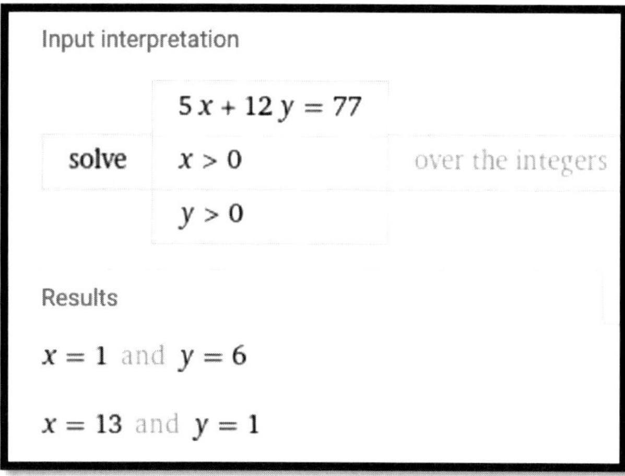

Figure 4.16. Solving the equation $5x + 12y = 77$ using Wolfram Alpha.

Remark 4.7. Similarly to what was said in Remark 4.6, the third scenario can be described through the recursive equation

$$g_{n+1} = 2g_n + g_{n-1}, g_1 = g_2 = 1. \tag{4.13}$$

Solving (4.13) using Wolfram Alpha yields $g_n = C_1(1 - \sqrt{2})^n + C_2(1 + \sqrt{2})^n$ from where, using initial values from (4.13), the tool can produce $C_1 = -\frac{1}{2} - \frac{1}{\sqrt{2}}$ and $C_2 = -\frac{1}{2} + \frac{1}{\sqrt{2}}$, thereby allowing for the solution of (4.13) to be written in the form

$$g_n = (-\tfrac{1}{2} - \tfrac{1}{\sqrt{2}})(1 - \sqrt{2})^n + (-\tfrac{1}{2} + \tfrac{1}{\sqrt{2}})(1 + \sqrt{2})^n. \tag{4.14}$$

Relation (4.14) can be confirmed by Maple as shown in Figure 4.17. Furthermore, one can show that both (4.14) and (4.12) represent, symbolically different, expressions for the number of cookies on the n-th plate in the context of the third scenario. Relation (4.12) depends on the initial values x and y, which were set $x = y = 1$ to get (4.14) from (4.13). In the spirit of computational triangulation, one can set $x = y = 1$ in (4.12) as well to get the expression

$$(-1 - \frac{3\sqrt{2}}{4})(1 - \sqrt{2})^n + (-1 + \frac{3\sqrt{2}}{4})(1 + \sqrt{2})^n +$$

$$(-1-\tfrac{3\sqrt{2}}{4})(1-\sqrt{2})^{n+1}+(-1+\tfrac{3\sqrt{2}}{4})(1+\sqrt{2})^{n+1}, \tag{4.15}$$

alternatively representing the number of cookies on the n-th plate and then, as a computational triangulation of the second order, use Maple (Figure 4.18) to show that (4.15) and the right-hand side of (4.14) are identical. This concludes TITE activities of this chapter.

> $rsolve(\{g(n+1)=2\cdot g(n)+g(n-1), g(1)=1, g(2)=1\}, \{g\})$

$$\left\{g(n)=\left(-\frac{\sqrt{2}}{2}-\frac{1}{2}\right)(-\sqrt{2}+1)^n+\left(-\frac{1}{2}+\frac{\sqrt{2}}{2}\right)(1+\sqrt{2})^n\right\}$$

Figure 4.17. Solving recursive equation (4.13) using Maple.

> $R(n) := \left(-\frac{1}{2}-\frac{1}{\sqrt{2}}\right)\cdot(1-\sqrt{2})^n+\left(-\frac{1}{2}+\frac{1}{\sqrt{2}}\right)\cdot(1+\sqrt{2})^n+\left(1+3\cdot\frac{\sqrt{2}}{4}\right)\cdot(1-\sqrt{2})^n+\left(1-3\cdot\frac{\sqrt{2}}{4}\right)\cdot(1+\sqrt{2})^n+\left(1+3\cdot\frac{\sqrt{2}}{4}\right)\cdot(1-\sqrt{2})^{n+1}+\left(1-3\cdot\frac{\sqrt{2}}{4}\right)\cdot(1+\sqrt{2})^{n+1}$

$$R := n \mapsto \left(-\frac{1}{2}-\frac{1}{\sqrt{2}}\right)\cdot(1-\sqrt{2})^n+\left(-\frac{1}{2}+\frac{1}{\sqrt{2}}\right)\cdot(1+\sqrt{2})^n+\left(1+\frac{3\cdot\sqrt{2}}{4}\right)\cdot(1-\sqrt{2})^n+\left(1-\frac{3\cdot\sqrt{2}}{4}\right)\cdot(1+\sqrt{2})^n+\left(1+\frac{3\cdot\sqrt{2}}{4}\right)\cdot(1-\sqrt{2})^{n+1}+\left(1-\frac{3\cdot\sqrt{2}}{4}\right)\cdot(1+\sqrt{2})^{n+1}$$

> $simplify(R(n))$

$$0$$

Figure 4.18. Showing that formulas (4.14) and (4.15) are identical.

Conclusion

The chapter, motivated by the primary school context, explored the ideas of computational triangulation using three scenarios grounded in contextually simple but conceptually intricate recursive rules that governed the placement of cookies on plates. It turned out that simple recursive rules generate several number sequences associated with the names of famous contributors to mathematical knowledgebase. The use of different digital tools demonstrated triangulation between methods, an approach to sociology aimed at acquiring rigor in scholarship. The scope of this chapter is appropriate to be included in problem-solving courses for prospective teachers of multiple grades. The next chapter will consider the above three scenarios by allowing the quantities of cookies placed on plates to be rational numbers.

Chapter 5

Computational Triangulation with Cookies on Plates Extended to Fractions

5.1. Introduction

The activities with cookies on plates in the context of the first scenario (Chapter 4, Section 4.2) were used with a class of 3rd graders who were given two-sided counters as a substitute for cookies, requiring the number of cookies on the last (fourth or fifth) plate to be a given whole number. The goal of the activities was to investigate how children can see patterns formed by cookies-counters and describe the numeric patterns found in the string of four or five integers (Abramovich & Griffin, 2025). An unexpected outcome of the activities was the display of collateral creativity (Abramovich & Freiman, 2023) by a 3rd grader who, through play with counters, used his own representation of the half of a cookie, by putting a counter in the vertical position using its thick edge as the base. Indeed, such unexpected representation of a part of a cookie was in the true spirit of collateral creativity understood as mostly accidental but nonetheless educationally gratifying outcome of using technology, both physical and digital. This outcome made it possible to recognize the potential of the student's collateral creativity for extending activities of Chapter 4 to fractions. It is the playful context of cookies on plates within which collateral creativity of the student was displayed that made such extension possible for all three scenarios that were discussed in the previous chapter.

5.2. The First Scenario of Cookies on Plates Extended to Fractions

The primary school context of Chapter 4 can be extended to include middle school mathematics if at least one of the initial values of Fibonacci-like numbers is allowed to be a reciprocal of an integer, that is, a unit fraction. Such extension means that pieces of cookies expressed by different mixed fractions (like $1\frac{1}{2}$) may appear on the plates. For example, in the case of the first scenario (when the number of cookies beginning from the 3rd plate is the sum of the number of cookies on the previous two plates), if the 1st plate has half a cookie and the 2nd plate has one cookie, then the following sequence would represent the number of cookies on the first eight plates: $\frac{1}{2}, 1, 1\frac{1}{2}, 2\frac{1}{2}, 4, 6\frac{1}{2}, 10\frac{1}{2}, 17$. Indeed, the equations $2x + 3y = 4$ and $8x + 13y = 17$ (see sequence (4.1), Chapter 4), representing 4 and 17 cookies on the 5th and 8th plates, respectively, are satisfied for $x = \frac{1}{2}$ and $y = 1$. Because the sum of two irreducible fractions with denominator 2 is an integer, between two plates with an integer number of cookies there are always two plates having half of a cookie. Put another way, when the first plate has half a cookie and the second plate has only full cookies, the transient time (length of transition) between two plates without half a cookie can be measured by two plates. This raises a question of finding the transient time between two plates with the whole number of cookies when the first plate has 1/3, 1/4, 1/5, and so on of a cookie, and the second plate has only full cookies. Is there a pattern formed by the corresponding transition times? In the spirit of computational triangulation, this question will be addressed below using different digital tools.

Also, if in the last two linear equations both x and y may be fractions, many solutions exist. For example, setting $y = \frac{1}{2}$ yields either $x = \frac{5}{4}$ (for the 5th plate) or $x = \frac{21}{16}$ (for the 8th plate) as $2 \cdot \frac{5}{4} + 3 \cdot \frac{1}{2} = 4$ and $8 \cdot \frac{21}{16} + 13 \cdot \frac{1}{2} = 17$, respectively. Likewise, setting $y = \frac{1}{3}$ yields either $x = \frac{3}{2}$ (for the 5th plate) or $x = \frac{19}{12}$ (for the 8th plate) as $2 \cdot \frac{3}{2} + 3 \cdot \frac{1}{3} = 4$ and $8 \cdot \frac{19}{12} + 13 \cdot \frac{1}{3} = 17$.

Recursive relation (4.2) of Chapter 4 representing the first scenario can be modified to the form

$$a_{n+1} = a_n + a_{n-1}, a_1 = \frac{1}{k}, a_2 = 1, \tag{5.1}$$

where integer $k \geq 2$, and solved by Maple (Figure 5.1) to get

$$a_n(k) = \frac{(k(\sqrt{5}+5)-3\sqrt{5}-5)(\frac{-\sqrt{5}+1}{2})^n+(k(-\sqrt{5}+5)+3\sqrt{5}-5)(\frac{\sqrt{5}+1}{2})^n}{10k} \tag{5.2}$$

When $k = 2$, the value of $a_n(k)$ represented by (5.2) has the form

$$a_n(2) = \frac{(5+\sqrt{5} - \frac{3\sqrt{5}}{2} - \frac{5}{2})(\frac{1-\sqrt{5}}{2})^n}{10} + \frac{(5-\sqrt{5} + \frac{3\sqrt{5}}{2} - \frac{5}{2})(\frac{1+\sqrt{5}}{2})^n}{10}$$

which, after a paper-and-pencil (i.e., a TI) simplification, is modeled (a TE activity) by Wolfram Alpha for the first 30 plates in Figure 5.2.

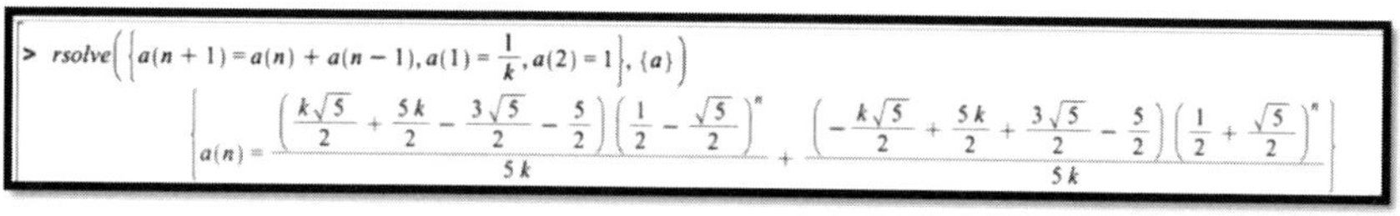

```
> rsolve({a(n+1) = a(n) + a(n-1), a(1) = 1/k, a(2) = 1}, {a})
```

$$\left\{ a(n) = \frac{\left(\frac{k\sqrt{5}}{2} + \frac{5k}{2} - \frac{3\sqrt{5}}{2} - \frac{5}{2}\right)\left(\frac{1}{2} - \frac{\sqrt{5}}{2}\right)^n}{5k} + \frac{\left(-\frac{k\sqrt{5}}{2} + \frac{5k}{2} + \frac{3\sqrt{5}}{2} - \frac{5}{2}\right)\left(\frac{1}{2} + \frac{\sqrt{5}}{2}\right)^n}{5k} \right\}$$

Figure 5.1. Using Maple to solve recursive relation (5.1).

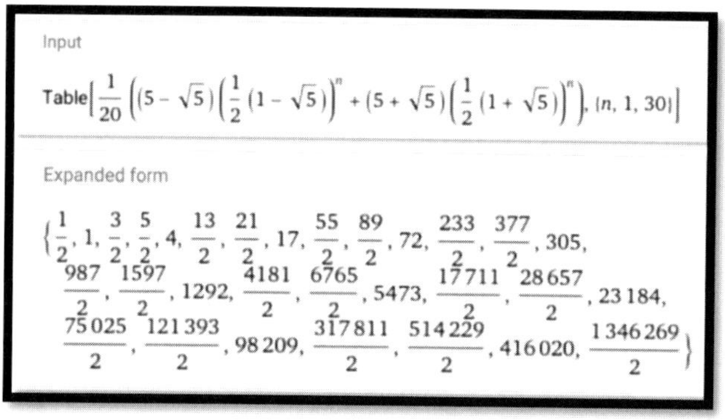

Input

$$\text{Table}\left[\frac{1}{20}\left((5-\sqrt{5})\left(\frac{1}{2}(1-\sqrt{5})\right)^n + (5+\sqrt{5})\left(\frac{1}{2}(1+\sqrt{5})\right)^n\right), \{n, 1, 30\}\right]$$

Expanded form

$$\left\{\frac{1}{2}, 1, \frac{3}{2}, \frac{5}{2}, 4, \frac{13}{2}, \frac{21}{2}, 17, \frac{55}{2}, \frac{89}{2}, 72, \frac{233}{2}, \frac{377}{2}, 305, \frac{987}{2}, \frac{1597}{2}, 1292, \frac{4181}{2}, \frac{6765}{2}, 5473, \frac{17711}{2}, \frac{28657}{2}, 23184, \frac{75025}{2}, \frac{121393}{2}, 98209, \frac{317811}{2}, \frac{514229}{2}, 416020, \frac{1346269}{2}\right\}$$

Figure 5.2. Wolfram Alpha generates the number of cookies starting with 1/2 and 1.

Alternatively, Wolfram Alpha (Figure 5.3) provides the following solution to recursive equation (5.1)

$$a_n(k) = \frac{(3-k)F_n + (k-1)L_n}{2k}$$

where $F_n = \frac{1}{\sqrt{5}}[(\frac{1+\sqrt{5}}{2})^n - (\frac{1-\sqrt{5}}{2})^n]$ and $L_n = (\frac{1+\sqrt{5}}{2})^n + (\frac{1-\sqrt{5}}{2})^n$ are, respectively, the n-th Fibonacci and Lucas numbers. Modifying (5.2) in a TI fashion to the form

$$a_n(k) = \frac{3-k}{2k} \cdot \frac{1}{\sqrt{5}}[(\frac{1+\sqrt{5}}{2})^n - (\frac{1-\sqrt{5}}{2})^n] + \frac{k-1}{2k}[(\frac{1+\sqrt{5}}{2})^n + \left(\frac{1-\sqrt{5}}{2}\right)^n\Big] = \frac{(3-k)F_n + (k-1)L_n}{2k}$$

shows that the two solutions to recursive equation (5.1) generated by Maple (Figure 5.1) and Wolfram Alpha (Figure 5.3) are identical.

Input
solve
$a(n+1) = a(n-1) + a(n)$
$a(1) = \frac{1}{k}$
$a(2) = 1$
for
$a(n)$
Result
$a(n) = \frac{-k F_n + 3 F_n + k L_n - L_n}{2k}$

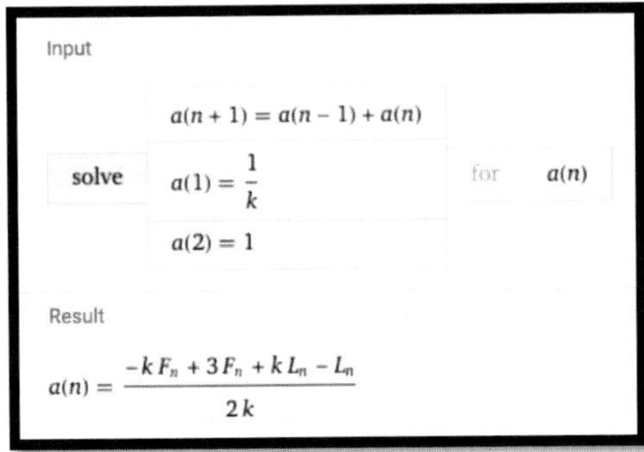

Figure 5.3. Solution of equation (5.1) generated by Wolfram Alpha.

The input data displayed at the top of Figure 5.2 is taken from the Maple's result of Figure 5.1 when $k = 2$. When $k = 3$, 4, and 5 we have (in a TI fashion) the following three sequences, respectively,

$$\frac{1}{3}, 1, 1\frac{1}{3}, 2\frac{1}{3}, 3\frac{2}{3}, 6;\ \frac{1}{4}, 1, 1\frac{1}{4}, 2\frac{1}{4}, 3\frac{1}{2}, 5\frac{3}{4}, 9\frac{1}{4}, 15;\ \frac{1}{5}, 1, 1\frac{1}{5}, 2\frac{1}{5}, 3\frac{2}{5}, 5\frac{3}{5}, 9.$$

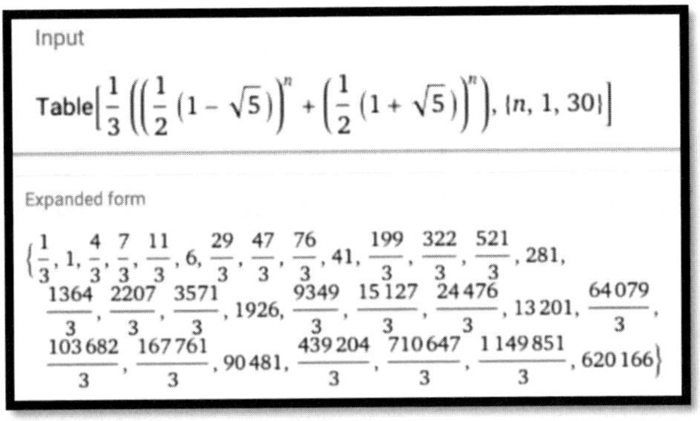

Input
$\text{Table}\left[\frac{1}{3}\left(\left(\frac{1}{2}(1-\sqrt{5})\right)^n + \left(\frac{1}{2}(1+\sqrt{5})\right)^n\right), \{n, 1, 30\}\right]$
Expanded form
$\{\frac{1}{3}, 1, \frac{4}{3}, \frac{7}{3}, \frac{11}{3}, 6, \frac{29}{3}, \frac{47}{3}, \frac{76}{3}, 41, \frac{199}{3}, \frac{322}{3}, \frac{521}{3}, 281,$
$\frac{1364}{3}, \frac{2207}{3}, \frac{3571}{3}, 1926, \frac{9349}{3}, \frac{15127}{3}, \frac{24476}{3}, 13201, \frac{64079}{3},$
$\frac{103682}{3}, \frac{167761}{3}, 90481, \frac{439204}{3}, \frac{710647}{3}, \frac{1149851}{3}, 620166\}$

Figure 5.4. Wolfram Alpha generates the number of cookies starting with 1/3 and 1.

The first sequence is confirmed by Wolfram Alpha in Figure 5.4; another two sequences will be confirmed below by a spreadsheet (Figure 5.7, rows 5 and 6). That is, in the context of the first scenario (Chapter 4, Section 4.2), when the 2nd plate has one cookie, starting with half a cookie on the 1st plate, the 5th plate will be the first (after the 2nd plate) to have a whole number of cookies (4); starting with one-third of a cookie on the 1st plate, the 6th plate will be the first (after the 2nd plate) to have a whole number of cookies (6); starting with one-fourth of a cookie on the first plate, the 8th plate will be the first (after the 2nd plate) to have a whole number of cookies (15); starting with one-fifth of a cookie on the 1st plate, the 7th plate will be the first (after the 2nd plate) to have a whole number of cookies (9). The transient times (measured by plates) between two plates with a whole number of cookies are: two plates in the case of 1/2, three plates in the case of 1/3, five plates in the case of 1/4, and four plates in the case of 1/5.

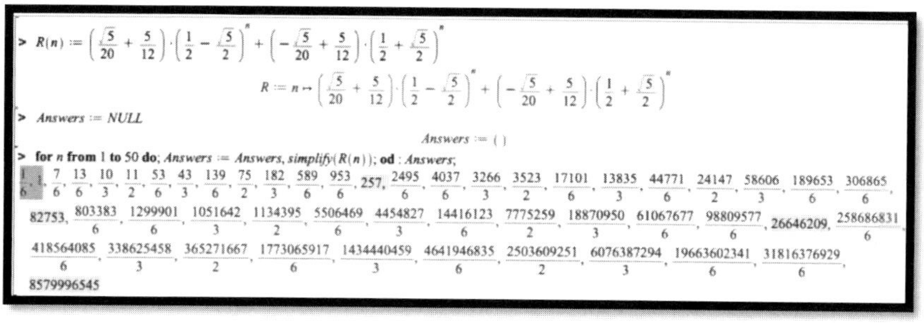

> $R(n) := \left(\frac{\sqrt{5}}{20}+\frac{5}{12}\right)\cdot\left(\frac{1}{2}-\frac{\sqrt{5}}{2}\right)^n+\left(-\frac{\sqrt{5}}{20}+\frac{5}{12}\right)\cdot\left(\frac{1}{2}+\frac{\sqrt{5}}{2}\right)^n$

$$R := n \mapsto \left(\frac{\sqrt{5}}{20}+\frac{5}{12}\right)\cdot\left(\frac{1}{2}-\frac{\sqrt{5}}{2}\right)^n+\left(-\frac{\sqrt{5}}{20}+\frac{5}{12}\right)\cdot\left(\frac{1}{2}+\frac{\sqrt{5}}{2}\right)^n$$

> *Answers* := *NULL*

$$Answers := (\ )$$

> **for** *n* **from** 1 **to** 50 **do**; *Answers* := *Answers*, *simplify*(*R*(*n*)); **od** : *Answers*;

$\frac{1}{6}, 1, \frac{7}{6}, \frac{13}{6}, \frac{10}{3}, \frac{11}{2}, \frac{53}{6}, \frac{43}{3}, \frac{139}{6}, \frac{75}{2}, \frac{182}{3}, \frac{589}{6}, \frac{953}{6}, 257, \frac{2495}{6}, \frac{4037}{6}, \frac{3266}{3}, \frac{3523}{2}, \frac{17101}{6}, \frac{13835}{3}, \frac{44771}{6}, \frac{24147}{2}, \frac{58606}{3}, \frac{189653}{6}, \frac{306865}{6}, 82753, \frac{803383}{6}, \frac{1299901}{6}, \frac{1051642}{3}, \frac{1134395}{2}, \frac{5506469}{6}, \frac{4454827}{3}, \frac{14416123}{6}, \frac{7775259}{2}, \frac{18870950}{3}, \frac{61067677}{6}, \frac{98809577}{6}, 26646209, \frac{258686831}{6}, \frac{418564085}{6}, \frac{338625458}{3}, \frac{365271667}{2}, \frac{1773065917}{6}, \frac{1434440459}{3}, \frac{4641946835}{6}, \frac{2503609251}{2}, \frac{6076387294}{3}, \frac{19663602341}{6}, \frac{31816376929}{6}, 8579996545$

Figure 5.5. Maple generates the number of cookies starting with 1/6 and 1.

Input

$$\text{Table}\left[\left(\frac{\sqrt{5}}{20}+\frac{5}{12}\right)\left(\frac{1}{2}-\frac{\sqrt{5}}{2}\right)^n+\left(-\frac{\sqrt{5}}{20}+\frac{5}{12}\right)\left(\frac{1}{2}+\frac{\sqrt{5}}{2}\right)^n, \{n, 1, 50\}\right]$$

Expanded form

$$\left\{\frac{1}{6}, 1, \frac{7}{6}, \frac{13}{6}, \frac{10}{3}, \frac{11}{2}, \frac{53}{6}, \frac{43}{3}, \frac{139}{6}, \frac{75}{2}, \frac{182}{3}, \frac{589}{6}, \frac{953}{6}, 257, \frac{2495}{6}, \frac{4037}{6}, \frac{3266}{3}, \frac{3523}{2}, \frac{17101}{6}, \frac{13835}{3}, \frac{44771}{6}, \frac{24147}{2}, \frac{58606}{3}, \frac{189653}{6}, \frac{306865}{6}, 82753, \frac{803383}{6}, \frac{1299901}{6}, \frac{1051642}{3}, \frac{1134395}{2}, \frac{5506469}{6}, \frac{4454827}{3}, \frac{14416123}{6}, \frac{7775259}{2}, \frac{18870950}{3}, \frac{61067677}{6}, \frac{98809577}{6}, 26646209, \frac{258686831}{6}, \frac{418564085}{6}, \frac{338625458}{3}, \frac{365271667}{2}, \frac{1773065917}{6}, \frac{1434440459}{3}, \frac{4641946835}{6}, \frac{2503609251}{2}, \frac{6076387294}{3}, \frac{19663602341}{6}, \frac{31816376929}{6}, 8579996545\right\}$$

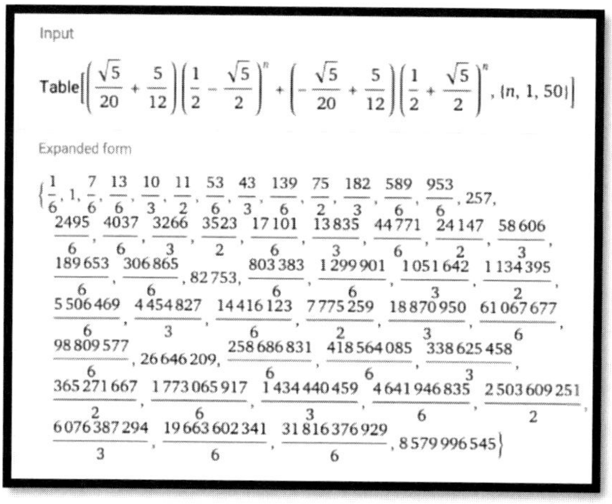

Figure 5.6. Wolfram Alpha generates the number of cookies starting with 1/6 and 1.

D3 f_x =((SQRT(5)*k*(1+SQRT(5))-SQRT(5)*(3+SQRT(5)))*((1-SQRT(5))/2)^n+(SQRT(5)*k*(-1+SQRT(5))-SQRT(5)*(3-SQRT(5)))*((1+SQRT(5))/2)^n)/(10*k)

k\n	1	2	3	4	5	6	7	8	9	10	11	12	13	14	15	16	17	18	19	20
1	1	1	2	3	5	8	13	21	34	55	89	144	233	377	610	987	1597	2584	4181	6765
2	1/2	1	1 1/2	2 1/2	4	6 1/2	10 1/2	17	27 1/2	44 1/2	72	116 1/2	188 1/2	305	493 1/2	798 1/2	1292	2090 1/2	3382 1/2	5473
3	1/3	1	1 1/3	2 1/3	3 2/3	6	9 2/3	15 2/3	25 1/3	41	66 1/3	107 1/3	173 2/3	281	454 2/3	735 2/3	1190 1/3	1926	3116 1/3	5042 1/3
4	1/4	1	1 1/4	2 1/4	3 1/2	5 3/4	9 1/4	15	24 1/4	39 1/4	63 1/2	102 3/4	166 1/4	269	435 1/4	704 1/4	1139 1/2	1843 3/4	2983 1/4	4827
5	1/5	1	1 1/5	2 1/5	3 2/5	5 3/5	9	14 3/5	23 3/5	38 1/5	61 4/5	100	161 4/5	261 4/5	423 3/5	685 2/5	1109	1794 2/5	2903 2/5	4697 4/5
6	1/6	1	1 1/6	2 1/6	3 1/3	5 1/2	8 5/6	14 1/3	23 1/6	37 1/2	60 2/3	98 1/6	158 5/6	257	415 5/6	672 5/6	1088 2/3	1761 1/2	2850 1/6	4611 2/3
7	1/7	1	1 1/7	2 1/7	3 2/7	5 3/7	8 5/7	14 1/7	22 6/7	37	59 6/7	96 6/7	156 5/7	253 4/7	410 2/7	663 6/7	1074 1/7	1738	2812 1/7	4550 1/7
8	1/8	1	1 1/8	2 1/8	3 1/4	5 3/8	8 5/8	14	22 5/8	36 5/8	59 1/4	95 7/8	155 1/8	251	406 1/8	657 1/8	1063 1/4	1720 3/8	2783 5/8	4504
9	1/9	1	1 1/9	2 1/9	3 2/9	5 1/3	8 5/9	13 8/9	22 4/9	36 1/3	58 7/9	95 1/9	153 8/9	249	402 8/9	651 8/9	1054 7/9	1706 2/3	2761 4/9	4468 1/9
10	1/10	1	1 1/10	2 1/10	3 1/5	5 3/10	8 1/2	13 4/5	22 3/10	36 1/10	58 2/5	94 1/2	152 9/10	247 2/5	400 3/10	647 7/10	1048	1695 7/10	2743 7/10	4439 2/5

Figure 5.7. Using spreadsheet to model formula (5.2) for $k = 1$ to 10, $n = 1$ to 20.

Moreover, showing no pattern, having one-sixth of a cookie on the 1st plate and a single cookie on the 2nd plate, the 14th plate will be the first one (after the 2nd plate) to have a whole number of cookies (257) with the transient time between two plates having a whole number of cookies equal eleven plates (see computational triangulation provided by Maple in Figure 5.5, by Wolfram Alpha in Figure 5.6, by a spreadsheet in Figure 5.7 (row 7), and by the Graphing Calculator in Figure 5.8). The computationally triangulated results are collected in the chart of Figure 5.9. One can see that there is no apparent pattern in the development of cookies on plates as the value of the unit fraction associated with the 1st plate decreases.

A	$B = \left(\frac{5}{12} + \frac{\sqrt{5}}{20}\right)\left(\frac{1-\sqrt{5}}{2}\right)^A + \left(\frac{5}{12} - \frac{\sqrt{5}}{20}\right)\left(\frac{1+\sqrt{5}}{2}\right)^A$
1	0.166667
2	1
3	1.16667
4	2.16667
5	3.33333
6	5.5
7	8.83333
8	14.3333
9	23.1667
10	37.5
11	60.6667
12	98.1667
13	158.833
14	257

Figure 5.8. Graphing Calculator shows 257 cookies on 14th plate starting from 1/6 and 1.

Cookies on the first two plates of the first scenario (initial values)	The first plate of the first scenario with a whole number of cookies	The number of cookies on that plate	Transient time (counting plates) to reach a whole number of cookies
(1/2, 1)	5	4	2
(1/3, 1)	6	6	3
(1/4, 1)	8	15	5
(1/5, 1)	7	9	4
(1/6, 1)	14	257	11
(1/7, 1)	10	37	7
(1/8, 1)	8	14	5

Figure 5.9. The first scenario with a fraction of a cookie on the first plate.

Remark 5.1. One can inquire about mathematical meaning of the results shown in Figure 5.9 regarding the transient time (measured by plates) it takes the process to reach a plate with a whole number of cookies. Consider the case when the 1st plate has half a cookie. As shown by sequence (4.1), Chapter 4, the number of cookies develops through a sequence the first five terms of which are x, y, $x + y$, $x + 2y$, $2x + 3y$, so that if $x = 1/2$ and $y = 1$ then the first time (after the 2nd plate) a plate would have a whole number of cookies when $2x + 3y$ assumes a whole number value, 4. Between y and $2x + 3y$ there are two terms, $x + y$ and $x + 2y$, that do not assume whole number values for $(x, y) = (1/2, 1)$. As shown in sequence (4.1), Chapter 4, between $2x + 3y$ and $8x + 13y$ (another sum with an even coefficient in x) there are also two terms that do not assume whole number values. Because the coefficients in the linear combinations of x and y are consecutive Fibonacci numbers, note that among three such numbers, two numbers are odd and one number is even. Indeed, starting from the first three Fibonacci numbers with this property (i.e., 1, 1, 2), the fourth and the fifth Fibonacci numbers are odd as a sum of two numbers of different parity is odd. Consequently, the sixth number is even as the sum of numbers of the same parity. This process will continue. In general, the sums $F_n x + F_{n+1} y, F_{n+1} x + F_{n+2} y, F_{n+2} x + F_{n+3} y$ represent the number of cookies on three consecutive plates in the context of the first scenario, where F_n is the n-th Fibonacci number. Noting that $F_{n+2} = F_n + F_{n+1}$ and $F_{n+3} = F_{n+1} + F_{n+2}$, these three linear combinations of x and y can be written as follows: $F_n x + F_{n+1} y, F_{n+1} x + (F_n + F_{n+1}) y$ and $(F_n + F_{n+1}) x + (F_{n+1} + F_{n+2}) y$. If F_n and F_{n+1} are odd numbers, then F_{n+2} is an even number; if F_n and F_{n+1} are numbers of different parity, then F_{n+2} is an odd number and F_{n+3} is either odd or even.

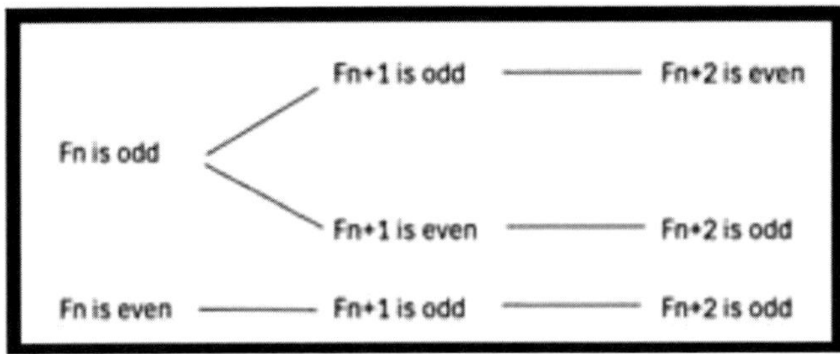

Figure 5.10. The parities of three consecutive Fibonacci numbers.

The tree diagram of Figure 5.10 represents the parities among three consecutive Fibonacci numbers. That is, a Fibonacci number has two chances to be odd and one chance to be even. The following parity relationships are possible among three consecutive Fibonacci numbers: (odd, odd, even), (odd, even, odd), (even, odd, odd). This explains why the transient time between two plates with a whole number of cookies in the case of $x = 1/2$ and $y = 1$ is measured by two plates.

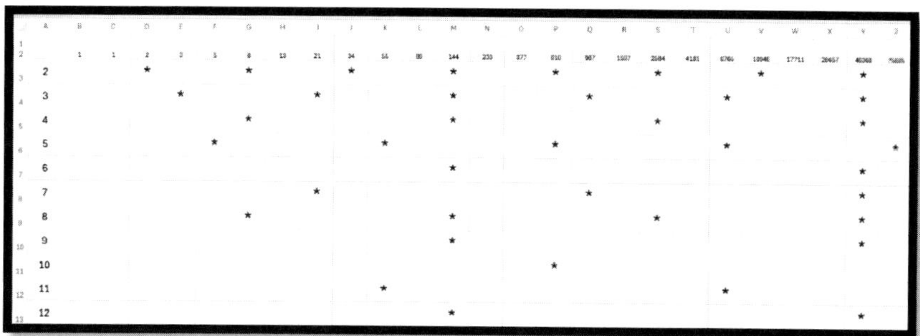

A	B	C	D	E	F	G	H	I	J	K	L	M	N	O	P	Q	R	S	T	U	V	W	X	Y	Z
	1	1	2	3	5	8	13	21	34	55	89	144	233	377	610	987	1597	2584	4181	6765	10946	17711	28657	46368	75025
2			★			★			★			★			★			★			★			★	
3				★				★				★				★				★				★	
4						★						★						★						★	
5					★					★					★					★					★
6												★												★	
7								★								★								★	
8						★						★						★						★	
9												★												★	
10															★										
11										★										★					
12												★												★	

Figure 5.11. Transient times of cookies on plates (the first scenario) shown by a spreadsheet.

Input

Table[F_n mod k, {k, 2, 7}, {n, 1, 50}]

Result

{{1, 1, 0, 1, 1, 0, 1, 1, 0, 1, 1, 0, 1, 1, 0, 1, 1, 0, 1, 1, 0, 1, 1, 0,
1, 1, 0, 1, 1, 0, 1, 1, 0, 1, 1, 0, 1, 1, 0, 1, 1, 0, 1, 1, 0, 1, 1, 0, 1, 1},
{1, 1, 2, 0, 2, 2, 1, 0, 1, 1, 2, 0, 2, 2, 1, 0, 1, 1, 2, 0, 2, 2, 1, 0, 1, 1,
2, 0, 2, 2, 1, 0, 1, 1, 2, 0, 2, 2, 1, 0, 1, 1, 2, 0, 2, 2, 1, 0, 1, 1},
{1, 1, 2, 3, 1, 0, 1, 1, 2, 3, 1, 0, 1, 1, 2, 3, 1, 0, 1, 1, 2, 3, 1, 0, 1, 1,
2, 3, 1, 0, 1, 1, 2, 3, 1, 0, 1, 1, 2, 3, 1, 0, 1, 1, 2, 3, 1, 0, 1, 1},
{1, 1, 2, 3, 0, 3, 3, 1, 4, 0, 4, 4, 3, 2, 0, 2, 2, 4, 1, 0, 1, 1, 2, 3, 0, 3,
3, 1, 4, 0, 4, 4, 3, 2, 0, 2, 2, 4, 1, 0, 1, 1, 2, 3, 0, 3, 3, 1, 4, 0},
{1, 1, 2, 3, 5, 2, 1, 3, 4, 1, 5, 0, 5, 5, 4, 3, 1, 4, 5, 3, 2, 5, 1, 0, 1, 1,
2, 3, 5, 2, 1, 3, 4, 1, 5, 0, 5, 5, 4, 3, 1, 4, 5, 3, 2, 5, 1, 0, 1, 1},
{1, 1, 2, 3, 5, 1, 6, 0, 6, 6, 5, 4, 2, 6, 1, 0, 1, 1, 2, 3, 5, 1, 6, 0, 6, 6,
5, 4, 2, 6, 1, 0, 1, 1, 2, 3, 5, 1, 6, 0, 6, 6, 5, 4, 2, 6, 1, 0, 1, 1}}

Figure 5.12. Remainders of the first 50 Fibonacci numbers when divided by k, $2 \le k \le 7$.

Remark 5.2. When $x = 1/k$, $k > 2$, and $y = 1$, one can use a spreadsheet (Figure 5.11) and Wolfram Alpha (Figure 5.12) to demonstrate the transient time phenomenon displayed in the chart of Figure 5.9. By changing k from 2 to 7, one can have Wolfram Alpha (Figure 5.12) to generate six 50-number strings, each string representing remainders of the first 50 Fibonacci numbers when divided by k, $2 \leq k \leq 7$. When $k = 2$, zeroes are separated by two non-zeroes; when $k = 3$, zeroes are separated by three non-zeroes; when $k = 4$, zeroes are separated by five non-zeroes; when $k = 5$, zeroes are separated by four non-zeroes; when $k = 6$, zeroes are separated by eleven non-zeroes; when $k = 7$, zeroes are separated by seven non-zeroes. In the spirit of computational triangulation, this result provided by Wolfram Alpha is consistent with what was shown by the spreadsheet of Figure 5.11 in which the asterisks represent zeroes separated by empty cells representing non-zeroes.

5.3. The Second Scenario of Cookies on Plates Extended to Fractions

In the case of the second scenario introduced in Chapter 4, Section 4.3, when the number of cookies on each plate beginning from the third one has twice as many cookies as the first plate of the previous pair plus the number of cookies on the second plate of the pair, the process of the development of cookies on plates being extended to fractions can be described by the recursive equation

$$a_{n+1} = a_n + 2a_{n-1}, a_1 = \frac{1}{k}, a_2 = 1 \quad (5.3)$$

the solution of which, found by Maple (Figure 5.13), is given by the formula

$$a_n(k) = \frac{(k-2)(-1)^n}{3k} + \frac{(k+1)2^n}{6k}. \quad (5.4)$$

Wolfram Alpha (Figure 5.14) provides almost identical form of solution to recursive equation (5.3), not requiring computational triangulation of the second order. Setting $k = 1$ in (5.4) as a special case (turning the second scenario into the first scenario) yields formula (4.9), Chapter 4, Section 4.3.

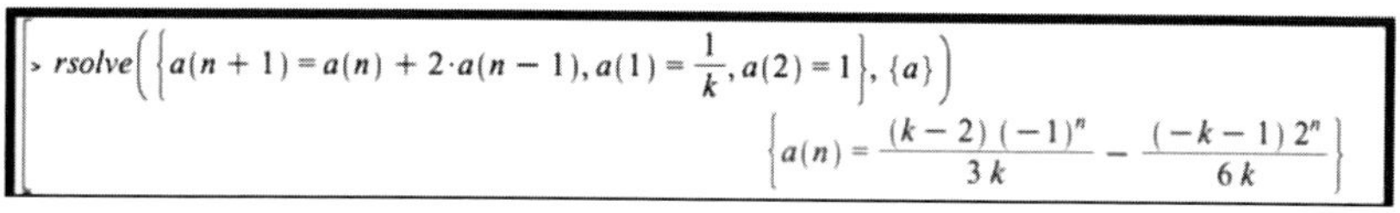

> $rsolve\left(\left\{a(n+1)=a(n)+2\cdot a(n-1), a(1)=\frac{1}{k}, a(2)=1\right\}, \{a\}\right)$

$$\left\{a(n)=\frac{(k-2)(-1)^n}{3k}-\frac{(-k-1)2^n}{6k}\right\}$$

Figure 5.13. Solving equation (5.3) using Maple.

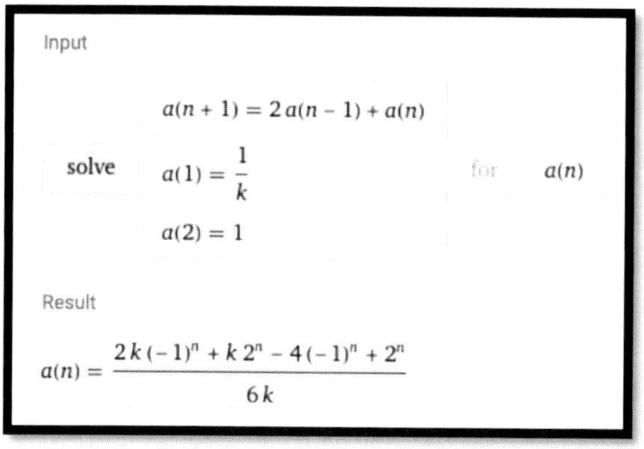

Input

solve $a(n+1) = 2a(n-1) + a(n)$, $a(1) = \frac{1}{k}$, $a(2) = 1$ for $a(n)$

Result

$$a(n) = \frac{2k(-1)^n + k\,2^n - 4(-1)^n + 2^n}{6k}$$

Figure 5.14. Solving equation (5.3) using Wolfram Alpha.

Remark 5.3. Setting $x = 1/2$ and $y = 1$ in formula (4.7), Chapter 4, Section 4.3, yields

$$f_n\left(\frac{1}{2},1\right) = \frac{2^n-4(-1)^n}{6}\cdot\frac{1}{2}+\frac{2^n+2(-1)^n}{6}\cdot 1 = \frac{3\cdot 2^n}{12} = 2^{n-2}.$$

Likewise, setting $k = 2$ in formula (5.4) yields $a_n(2) = \frac{3\cdot 2^n}{12} = \frac{2^n}{4} = 2^{n-2}$, thereby confirming results obtained by different tools (Maple, Wolfram Alpha, a spreadsheet), through different techniques (i.e., using formulas (4.6) and (5.4)), and explaining the presence of only whole number of cookies (powers of two) on every plate (see row 3 of the spreadsheet of Figure 5.15). Setting $x = 1/4$ and $y = 1$ in formula (4.7), Chapter 4, yields $f_n\left(\frac{1}{4},1\right) = \frac{2^n-4(-1)^n}{24}+\frac{2^n+2(-1)^n}{6} = \frac{5\cdot 2^n+4(-1)^n}{24} = \frac{10\cdot 2^{n-3}+(-1)^n}{6}$. Setting $k = 4$ in formula (5.4) yields $a_n(4) = \frac{2\cdot(-1)^n}{12}+\frac{5\cdot 2^n}{24} = \frac{(-1)^n}{6}+\frac{5\cdot 2^{n-2}}{6} = \frac{10\cdot 2^{n-3}+(-1)^n}{6}$. One can see that for $n \geq 3$ the expression $10\cdot 2^{n-3}+(-1)^n$ is one smaller or one greater than a power of ten, something that can (almost) never be divisible by six (except $n = 2$). This explains why the spreadsheet of Figure 5.15 does not display a whole number in row 5 (the case $k = 4$).

B2 f_x =(($A2-2)*(-1)^B$1)/(3*$A2) + (($A2+1)*2^B$1)/(6*$A2)

k\n	1	2	3	4	5	6	7	8	9	10
1	1	1	3	5	11	21	43	85	171	341
2	1/2	1	2	4	8	16	32	64	128	256
3	1/3	1	1 2/3	3 2/3	7	14 1/3	28 1/3	57	113 2/3	227 2/3
4	1/4	1	1 1/2	3 1/2	6 1/2	13 1/2	26 1/2	53 1/2	106 1/2	213 1/2
5	1/5	1	1 2/5	3 2/5	6 1/5	13	25 2/5	51 2/5	102 1/5	205
6	1/6	1	1 1/3	3 1/3	6	12 2/3	24 2/3	50	99 1/3	199 1/3
7	1/7	1	1 2/7	3 2/7	5 6/7	12 3/7	24 1/7	49	97 2/7	196 2/7
8	1/8	1	1 1/4	3 1/4	5 3/4	12 1/4	23 3/4	48 1/4	95 3/4	192 1/4
9	1/9	1	1 2/9	3 2/9	5 2/3	12 1/9	23 4/9	47 2/3	94 5/9	189 8/9
10	1/10	1	1 1/5	3 1/5	5 3/5	12	23 1/5	47 1/5	93 3/5	188

k\n	11	12	13	14	15	16	17	18	19	20
1	683	1365	2731	5461	10923	21845	43691	87381	174763	349525
2	512	1024	2048	4096	8192	16384	32768	65536	131072	262144
3	455	910 1/3	1820 1/3	3641	7281 2/3	14563 2/3	29127	58254 1/3	116508 1/3	233017
4	426 1/2	853 1/2	1706 1/2	3413 1/2	6826 1/2	13653 1/2	27306 1/2	54613 1/2	109226 1/2	218453 1/2
5	409 2/5	819 2/5	1638 1/5	3277	6553 2/5	13107 2/5	26214 1/5	52429	104857 2/5	209715 2/5
6	398	796 2/3	1592 2/3	3186	6371 1/3	12743 1/3	25486	50972 2/3	101944 2/3	203890
7	389 6/7	780 3/7	1560 1/7	3121	6241 2/7	12483 2/7	24965 6/7	49932 3/7	99864 1/7	199729
8	383 3/4	768 1/4	1535 3/4	3072 1/4	6143 3/4	12288 1/4	24575 3/4	49152 1/4	98303 3/4	196608 1/4
9	379	758 7/9	1516 7/9	3034 1/3	6067 8/9	12136 5/9	24272 1/3	48545 4/9	97090 1/9	194181
10	375 1/5	751 1/5	1501 3/5	3004	6007 1/5	12015 1/5	24029 3/5	48060	96119 1/5	192239 1/5

Figure 5.15. Using spreadsheet to model formula (5.4) for $k = 1$ to 10, $n = 1$ to 20.

Likewise, setting $k = 8$ in formula (5.4) yields $a_n(8) = \frac{6\cdot(-1)^n}{24} + \frac{9\cdot 2^n}{48} = \frac{3\cdot 2^{n-2}+(-1)^n}{4}$, the expression that is (almost) never divisible by 4 (except $n = 2$; see row 9 of the spreadsheet of Figure 5.15 free from integers except for cell C9).

Remark 5.4. Setting $k = 1$ in formula (5.4) yields the sequence $a_n(1) = \frac{(-1)^{n+1}+2^n}{3}$ the values of which are generated in row 2 of the spreadsheet of Figure 5.16. Beginning from row 3, the spreadsheet is programmed to generate an asterisk when the numbers displayed in row 2 are divisible by the corresponding number in column A. For example, one can see an asterisk in cell D4 meaning that a_3 (cell D2) is divisible by 3 (cell A4). An empty cell D4 means that a_4 (cell E2) is not divisible by 3 (cell A4). Also, the spreadsheet demonstrates that $a_n(k)$ defined by formula (5.4) is always an odd number as rows 3, 5, 7, 9, 11, and 13 (in which divisibility by the powers of 2 is tested) are asterisk-free. Furthermore, the absence of asterisks in row 11 indicates that the sequence $a_n(k)$ is not divisible by five as well. In order to prove that the expression $\frac{(-1)^{n+1}+2^n}{3}$ is an odd integer, one can construct the relation $\frac{(-1)^{n+1}+2^n}{3} = 2k + 1$ whence $k = \frac{(-1)^{n+1}+2^n-3}{6}$. It remains to be proved that the expression $P(n) = (-1)^{n+1} + 2^n - 3$ is divisible by six for all natural values of n. This can be proved by Maple-based mathematical induction (see the conclusion of Section 1.3.3, Chapter 1). Checking the base clause for $n = 1$ yields $P(1) = 0$ implying divisibility by six. To demonstrate the inductive transfer from n to $n + 1$, two cases must be considered: n is either an even or an odd number.

B2 fx =((-1)^(B$1+1)+2^B$1)/3

k\n	1	2	3	4	5	6	7	8	9	10	11	12	13	14	15	16	17	18	19	20	21
	1	1	3	5	11	21	43	85	171	341	683	1365	2731	5461	10923	21845	43691	87381	174763	349525	699051
2																					
3			*			*			*			*			*			*			*
4																					
5				*				*				*				*				*	
6																					
7						*						*						*			
8																					
9									*									*			
10																					
11					*					*					*					*	
12																					

Figure 5.16. Transient times of cookies on plates (the second scenario) shown by a spreadsheet.

```
> P(n) := 2^n + (-1)^(n+1) - 3
                                        P := n ↦ 2^n + (-1)^(n+1) - 3
> PE(n) := P(2·n)
                                        PE := n ↦ P(2·n)
> PE(n + 1) - PE(n)
                                        2^(2n+2) + (-1)^(2n+3) - 2^(2n) - (-1)^(2n+1)
> simplify(%)
                                        3 4^n
> PO(n) := P(2·n + 1)
                                        PO := n ↦ P(2·n + 1)
> PO(n + 1) - PO(n)
                                        2^(2n+3) + (-1)^(2n+4) - 2^(2n+1) - (-1)^(2n+2)
> simplify(%)
                                        6 4^n
```

Figure 5.17. Maple-based mathematical induction proof.

As shown in Figure 5.17, in both cases, the difference $P(n+1) - P(n)$ is divisible by six (as it is either $3 \cdot 4^n = 6 \cdot 2^{2n-1}$ or $6 \cdot 4^n$) and therefore, assuming $P(n)$ to possess this property, implies $P(n+1)$ is also divisible by six; that is, the transition from n to $n + 1$ holds true the divisibility of $P(n)$ by six .

5.4. The Third Scenario of Cookies on Plates Extended to Fractions

In the case of the third scenario (Chapter 4, Section 4.4), when the number of cookies on each plate beginning from the third one has twice as many cookies as the second plate of the previous pair plus the number of cookies on the first plate of the pair, the process of the development of cookies on plates being extended to fractions can be described by the recursive equation

$$a_{n+1} = 2a_n + a_{n-1}, a_1 = \frac{1}{k}, a_2 = 1 , \qquad (5.5)$$

the solution of which, found by Maple (Figure 5.18), is given by the formula

$$a_n(k) = \frac{(2k-4-(k-3)\sqrt{2})(1+\sqrt{2})^n}{4k} + \frac{(2k-4+(k-3)\sqrt{2})(1-\sqrt{2})^n}{4k}. \qquad (5.6)$$

$$> rsolve\left(\left\{a(n+1)=2\cdot a(n)+a(n-1), a(1)=\frac{1}{k}, a(2)=1\right\}, \{a\}\right)$$

$$\left\{a(n)=\frac{(2k-k\sqrt{2}-4+3\sqrt{2})(1+\sqrt{2})^n}{4k}+\frac{(k\sqrt{2}+2k-3\sqrt{2}-4)(-\sqrt{2}+1)^n}{4k}\right\}$$

Figure 5.18. Solving equation (5.5) using Maple.

Setting in formula (4.12), Chapter 4, $x = 1/2$ and $y = 1$ yields

$$\frac{1}{2}(-1-\frac{3\sqrt{2}}{4})(1-\sqrt{2})^n+\frac{1}{2}(-1+\frac{3\sqrt{2}}{4})(1+\sqrt{2})^n+(-1-\frac{3\sqrt{2}}{4})(1-\sqrt{2})^{n+1}+(-1+\frac{3\sqrt{2}}{4})(1+\sqrt{2})^{n+1}$$

$$=\frac{1}{2}(-1-\frac{3\sqrt{2}}{4})(3-2\sqrt{2})(1-\sqrt{2})^n+\frac{1}{2}(-1+\frac{3\sqrt{2}}{4})(3+2\sqrt{2})(1+\sqrt{2})^n$$

$$=\frac{(2\sqrt{2}-3)(4+3\sqrt{2})(1-\sqrt{2})^n+(3\sqrt{2}-4)(3+2\sqrt{2})(1+\sqrt{2})^n}{8}$$

$$=\frac{(8\sqrt{2}-12+12-9\sqrt{2})(1-\sqrt{2})^n+(9\sqrt{2}+12-12-8\sqrt{2})(1+\sqrt{2})^n}{8}$$

$$=\frac{-\sqrt{2}(1-\sqrt{2})^n+\sqrt{2}(1+\sqrt{2})^n}{8}=\frac{\sqrt{2}}{8}\left[\left(1+\sqrt{2}\right)^n-\left(1-\sqrt{2}\right)^n\right].$$

Likewise, setting $k = 2$ in formula (5.6) yields

$$a_n(2)=\frac{\sqrt{2}}{8}\left[\left(1+\sqrt{2}\right)^n-\left(1-\sqrt{2}\right)^n\right]. \quad (5.7)$$

Input

$$\text{Table}\left[\frac{\sqrt{2}}{8}\left(\left(1+\sqrt{2}\right)^n-\left(1-\sqrt{2}\right)^n\right), \{n, 1, 15\}\right]$$

Expanded form

$$\left\{\frac{1}{2}, 1, \frac{5}{2}, 6, \frac{29}{2}, 35, \frac{169}{2}, 204, \frac{985}{2}, 1189, \frac{5741}{2}, 6930, \frac{33461}{2}, 40391, \frac{195025}{2}\right\}$$

Figure 5.19. Modeling formula (5.7) using Wolfram Alpha.

	A	B	C	D	E	F	G	H	I	J	K	L	M	N	O	P
1	k\n	1	2	3	4	5	6	7	8	9	10	11	12	13	14	15
2	1	1	1	3	7	17	41	99	239	577	1393	3363	8119	19601	47321	114243
3	2	1/2	1	2 1/2	6	14 1/2	35	84 1/2	204	492 1/2	1189	2870 1/2	6930	16730 1/2	40391	97512 1/2
4	3	1/3	1	2 1/3	5 2/3	13 2/3	33	79 2/3	192 1/3	464 1/3	1121	2706 1/3	6533 2/3	15773 2/3	38081	91935 2/3
5	4	1/4	1	2 1/4	5 1/2	13 1/4	32	77 1/4	186 1/2	450 1/4	1087	2624 1/4	6335 1/2	15295 1/4	36926	89147 1/4
6	5	1/5	1	2 1/5	5 2/5	13	31 2/5	75 4/5	183	441 4/5	1066 3/5	2575	6216 3/5	15008 1/5	36233	87474 1/5
7	6	1/6	1	2 1/6	5 1/3	12 5/6	31	74 5/6	180 2/3	436 1/6	1053	2542 1/6	6137 1/3	14816 5/6	35771	86358 5/6
8	7	1/7	1	2 1/7	5 2/7	12 5/7	30 5/7	74 1/7	179	432 1/7	1043 2/7	2518 5/7	6080 5/7	14680 1/7	35441	85562 1/7
9	8	1/8	1	2 1/8	5 1/4	12 5/8	30 1/2	73 5/8	177 3/4	429 1/8	1036	2501 1/8	6038 1/4	14577 5/8	35193 1/2	84964 5/8
10	9	1/9	1	2 1/9	5 2/9	12 5/9	30 1/3	73 2/9	176 7/9	426 7/9	1030 1/3	2487 4/9	6005 2/9	14497 8/9	35001	84499 8/9
11	10	1/10	1	2 1/10	5 1/5	12 1/2	30 1/5	72 9/10	176	424 9/10	1025 4/5	2476 1/2	5978 4/5	14434 1/10	34847	84128 1/10

Figure 5.20. Using spreadsheet to model formula (5.6) for $k = 1$ to 10, $n = 1$ to 14.

Input

$$\text{Table}\left[\frac{(2k-4-(k-3)\sqrt{2})(1+\sqrt{2})^n}{4k}+\frac{(2k-4+(k-3)\sqrt{2})(1-\sqrt{2})^n}{4k},\ \{k,2,10\},\{n,1,15\}\right]$$

Expanded form

$$\begin{pmatrix}
\frac{1}{2} & 1 & \frac{5}{2} & 6 & \frac{29}{2} & 35 & \frac{169}{2} & 204 & \frac{985}{2} & 1189 & \frac{5741}{2} & 6930 & \frac{33461}{2} & 40391 & \frac{195025}{2} \\
\frac{1}{3} & 1 & \frac{7}{3} & \frac{17}{3} & \frac{41}{3} & 33 & \frac{239}{3} & \frac{577}{3} & \frac{1393}{3} & 1121 & \frac{8119}{3} & \frac{19601}{3} & \frac{47321}{3} & 38081 & \frac{275807}{3} \\
\frac{1}{4} & 1 & \frac{9}{4} & \frac{11}{2} & \frac{53}{4} & 32 & \frac{309}{4} & \frac{373}{2} & \frac{1801}{4} & 1087 & \frac{10497}{4} & \frac{12671}{2} & \frac{61181}{4} & 36926 & \frac{356589}{4} \\
\frac{1}{5} & 1 & \frac{11}{5} & \frac{27}{5} & 13 & \frac{157}{5} & \frac{379}{5} & 183 & \frac{2209}{5} & \frac{5333}{5} & 2575 & \frac{31083}{5} & \frac{75041}{5} & 36233 & \frac{437371}{5} \\
\frac{1}{6} & 1 & \frac{13}{6} & \frac{16}{3} & \frac{77}{6} & 31 & \frac{449}{6} & \frac{542}{3} & \frac{2617}{6} & 1053 & \frac{15253}{6} & \frac{18412}{3} & \frac{88901}{6} & 35771 & \frac{518153}{6} \\
\frac{1}{7} & 1 & \frac{15}{7} & \frac{37}{7} & \frac{89}{7} & \frac{215}{7} & \frac{519}{7} & 179 & \frac{3025}{7} & \frac{7303}{7} & \frac{17631}{7} & \frac{42565}{7} & \frac{102761}{7} & 35441 & \frac{598935}{7} \\
\frac{1}{8} & 1 & \frac{17}{8} & \frac{21}{4} & \frac{101}{8} & \frac{61}{2} & \frac{589}{8} & \frac{711}{4} & \frac{3433}{8} & 1036 & \frac{20009}{8} & \frac{24153}{4} & \frac{116621}{8} & \frac{70387}{2} & \frac{679717}{8} \\
\frac{1}{9} & 1 & \frac{19}{9} & \frac{47}{9} & \frac{113}{9} & \frac{91}{3} & \frac{659}{9} & \frac{1591}{9} & \frac{3841}{9} & \frac{3091}{3} & \frac{22387}{9} & \frac{54047}{9} & \frac{130481}{9} & 35001 & \frac{760499}{9} \\
\frac{1}{10} & 1 & \frac{21}{10} & \frac{26}{5} & \frac{25}{2} & \frac{151}{5} & \frac{729}{10} & 176 & \frac{4249}{10} & \frac{5129}{5} & \frac{4953}{2} & \frac{29894}{5} & \frac{144341}{10} & 34847 & \frac{841281}{10}
\end{pmatrix}$$

Figure 5.21. Modeling formula (5.6) using Wolfram Alpha.

One can see how different problem-solving techniques supported by Maple – solving recursive equation (4.11), Chapter 4, to get linear combinations (4.12) of the number of cookies on the first two plates, and solving recursive equation (5.5) in which the number of cookies on the first two plates served as the initial values – provide same result (in the former case by setting $x = 1/2, y = 1$; in the latter case setting $k = 2$), which, that is, formula (5.7), is treated then by Wolfram Alpha (Figure 5.19) to display the number of cookies on the first 15 plates. In the spirit of computational triangulation, the result shown in Figure 5.19 is confirmed by the spreadsheet of Figure 5.20 (row 3, $k = 2$) and Wolfram Alpha (Figure 5.21, the first row, $k = 2$) through modeling formula (5.6) – a solution to recursive equation (5.5). One can also see that the numbers in the range O3:O11 of the spreadsheet of Figure 5.20 coincide with the numbers in the penultimate column of the Wolfram Alpha table of Figure 5.21. This is exactly what computational triangulation as a way of providing rigor in the use of different digital tools means.

Conclusion

The chapter extended explorations of Chapter 4 to allow for the number of cookies to be represented by rational numbers within the same three scenarios. To this end, unit fractions were used to represent quantities of cookies put on the first plate. As a result, mixed fractions as well as integers represented the quantities of cookies appearing on the plates. This extension motivated using new focus of explorations dealing with the transient time needed to move from plate to plate having an integer number of cookies. Computational triangulation carried out in the context of different digital tools confirmed the results each tool provided that, however, demonstrated no pattern followed by the transient time as the initial values of recursive rules changed. In the context of the first scenario, the transient time was connected to the parity properties of Fibonacci numbers. The scope of this chapter is appropriate to be included in problem-solving courses for prospective teachers of multiple grades. The next chapter will return to mathematical inquiries into the context of Bowen's emotional systems discussed in Chapter 1.

Chapter 6

Computational Triangulation Using Secondary School Context

6.1. Introduction

As mentioned by Kline (1985), "A farmer who seeks the rectangle of maximum area with given perimeter might, after finding the answer to his question, turn to gardening, but a mathematician who obtains such a neat result would not stop there" (p. 133). This citation has several implications for mathematics teacher education. First, it shows what it might mean by expecting a teacher to act as a mathematician in the classroom (Conference Board of the Mathematical Sciences, 2012). Such an action on the part of a teacher requires seeing an extension of problem solving through the lens of problem posing (Cai & Leikin, 2025). The extension stems from one's ability to reflect on a problem solved by asking a "what if" question. In many cases, such questions can be asked by collaterally creative students and a teacher must take an intellectual risk to use unexpected questions as an opportunity for future didactic developments. Second, the citation is applicable to almost any situation that happens in a mathematics classroom. That is, teaching begins with a real-life situation which is then used as a springboard into relevant mathematics (Ontario Ministry of Education, 2020). Third, the citation shows how mathematics developed historically by its creators' recognizing neatness in results obtained through solving problems (Kafoussi & Margaritidou, 2023), something that shows directions for going through the cognitive cycle solve-reflect-pose (Abramovich, 2015; Otun & Njoku, 2020). These ideas will be used in this chapter for expanding the context of Bowen's emotional systems discussed in Chapter 1, Section 1.4.

6.2. Expanding Bowen's Emotional System by Forming Polygons

One can continue exploring the context of Bowen's emotional systems through the following mathematically-oriented query: How many convex quadrilaterals can be created by connecting four, five, six and so on individuals? Figure 6.1 shows how five individuals A, B, C, D and E can form five convex quadrilaterals ABCD, ABCE, ABDE, ACDE and BCDE. In other words, the number 5 is the number of combinations (without repetition) of four objects out of five different objects. That is, $C_5^4 = 5$. Likewise, $C_6^4 = 15$ is the number of convex quadrilaterals formed by a six-person system, $C_7^4 = 35$ is the number of convex quadrilaterals formed by a seven-person system, $C_8^4 = 40$ is the number of convex quadrilaterals formed by an eight-person system, and so on. Note that although the number of non-degenerate quadrilaterals (i.e., quadrilaterals preserving its traditional structure of not having more than two vertexes on the same line) formed by a five-person system is five regardless of whether only convex or both convex (Figure 6.1) and concave (Figure 6.2) quadrilaterals are considered, we restrict our consideration for convex polygons only (to be consistent with triangles which are always convex) formed by an m-person system, $m \geq 4$.

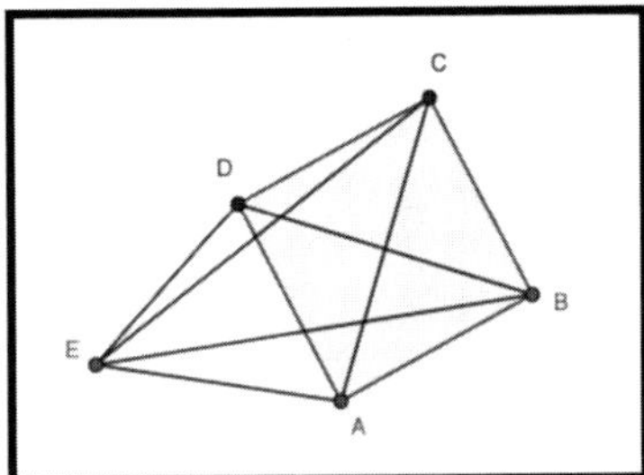

Figure 6.1. Five quadrilaterals in a five-person system.

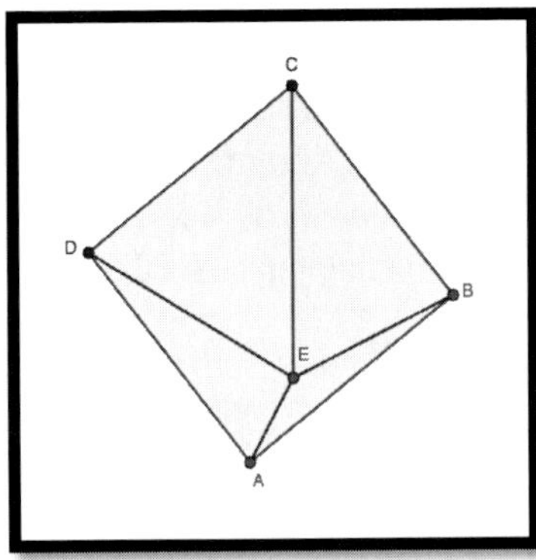

Figure 6.2. Five convex and concave quadrilaterals in a five-person system.

The notion of computational triangulation can be applied to develop a sequence representing the number of quadrilaterals formed by an m-person system, $m \geq 4$. One way to develop such a sequence is, by using Wolfram Alpha, to generate a numeric table of values of $C_m^4 = \frac{m!}{4!(m-4)!}$ for $m = 4, 5, \ldots, 10$ (Figure 6.3). Another way is to extend the spreadsheet of Figure 1.8, Chapter 1, by computing partial sums of tetrahedral numbers (Figure 6.4, row 4). The third way is to use Wolfram Alpha (Figure 6.5) to find the sum of the first m tetrahedral numbers. Finally, one can use a spreadsheet (Figure 6.6) to create an array of the numbers $C_m^k = \frac{m!}{k!(m-k)!}$ representing the number of convex polygons with k sides formed by connecting m points on the plane, $m \geq k$; in other words, an m-person system can form k-gonal emotional systems in $\frac{m!}{k!(m-k)!}$ ways. In particular, as shown in the spreadsheet of Figure 6.6 (cell F4), out of seven people, a five-person emotional system can be developed in 21 ways; in other words, one can create 21 convex pentagons by connecting seven points on the plane.

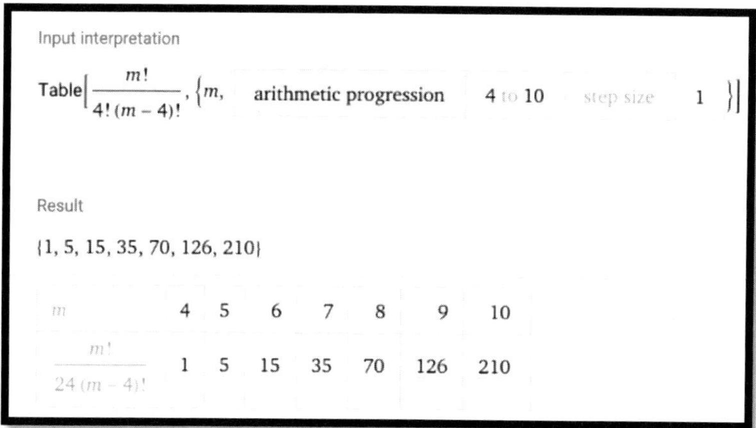

Figure 6.3. Selecting four objects out of m objects ($4 \leq m \leq 10$).

B4 | f_x =A4+B3

	A	B	C	D	E	F	G	H	I	J	K	L	M	N
1	1	2	3	4	5	6	7	8	9	10	11	12	13	14
2	1	3	6	10	15	21	28	36	45	55	66	78	91	105
3	1	4	10	20	35	56	84	120	165	220	286	364	455	560
4	1	5	15	35	70	126	210	330	495	715	1001	1365	1820	2380

Figure 6.4. Computing partial sums of tetrahedral numbers in row 4.

Sum

$$\sum_{i=1}^{m} \frac{1}{6}\, i\,(i+1)\,(i+2) = \frac{1}{24}\, m\,(m+1)\,(m+2)\,(m+3)$$

Figure 6.5. Finding general formula for the partial sums of tetrahedral numbers.

	A	B	C	D	E	F	G	H	I	J	K
1	k\m	3	4	5	6	7	8	9	10	11	12
2	3	1	4	10	20	35	56	84	120	165	220
3	4		1	5	15	35	70	126	210	330	495
4	5			1	6	21	56	126	252	462	792
5	6				1	7	28	84	210	462	924
6	7					1	8	36	120	330	792
7	8						1	9	45	165	495
8	9							1	10	55	220
9	10								1	11	66

Figure 6.6. Generating C_m^k, $m \geq k$, within a spreadsheet.

6.3. Connecting Different Integer Sequences

The formulas (see also Figures 1.9, 6.5, 6.7, 6.8)

$$\sum_{i=1}^{m} i = \frac{m(m+1)}{2} = \frac{\langle m\rangle_2}{2},$$

$$\sum_{i=1}^{m} \frac{i(i+1)}{2} = \frac{m(m+1)(m+2)}{6} = \frac{\langle m\rangle_3}{6},$$

$$\sum_{i=1}^{m} \frac{i(i+1)(i+2)}{6} = \frac{m(m+1)(m+2)(m+3)}{24} = \frac{\langle m\rangle_4}{24},$$

$$\sum_{i=1}^{m}\frac{i(i+1)(i+2)(i+3)}{24}=\frac{m(m+1)(m+2)(m+3)(m+4)}{120}=\frac{\langle m\rangle_5}{120},$$

where $\langle m\rangle_n = m(m+1)\dots(m+n-1)$ is the rising factorial function called, alternatively, in the theory of special functions, the Pochhammer symbol named after a German mathematician Leo August Pochhammer (1841-1920), representing partial sums of the numbers (binomial coefficients) located in rows 1, 2, 3, 4 of the spreadsheet of Figure 6.4, suggest the following multiplicatively simple yet conceptually profound identities:

$$m\times\frac{(m+1)}{2}=\frac{m(m+1)}{2}, \tag{6.1}$$

$$\frac{m(m+1)}{2}\times\frac{(m+2)}{3}=\frac{m(m+1)(m+2)}{6}, \tag{6.2}$$

$$\frac{m(m+1)(m+2)}{6}\times\frac{(m+3)}{4}=\frac{m(m+1)(m+2)(m+3)}{24}, \tag{6.3}$$

$$\frac{m(m+1)(m+2)(m+3)}{24}\times\frac{(m+4)}{5}=\frac{m(m+1)(m+2)(m+3)(m+4)}{120}. \tag{6.4}$$

Sum

$$\sum_{i=1}^{m}\frac{1}{24}\,i\,(i+1)\,(i+2)\,(i+3)=\frac{1}{120}\,m\,(m+1)\,(m+2)\,(m+3)\,(m+4)$$

Figure 6.7. Summation with the help of Wolfram Alpha.

Sum

$$\sum_{i=1}^{m}\frac{1}{120}\,i\,(i+1)\,(i+2)\,(i+3)\,(i+4)=$$
$$\frac{1}{720}\,m\,(m+1)\,(m+2)\,(m+3)\,(m+4)\,(m+5)$$

Figure 6.8. Summation with the help of Wolfram Alpha.

For example, identity (6.1) prompts seeing the following connection between natural and triangular numbers (rows 1 and 2, respectively, of the spreadsheet of Figure 6.4): $1\times\frac{1+1}{2}=1, 2\times\frac{2+1}{2}=3, 3\times\frac{3+1}{2}=6, 4\times\frac{4+1}{2}=10$, and so on. Likewise, identity (6.2) prompts seeing the following connection between triangular and tetrahedral numbers (rows 2 and 3, respectively, of the spreadsheet of Figure 6.4): $1\times\frac{1+2}{3}=1, 3\times\frac{2+2}{3}=4, 6\times\frac{3+2}{3}=10, 10\times\frac{4+2}{3}=20$, and so on. Similar connections between numbers in rows 3 and 4 as well in rows 4 and 5 of the spreadsheet of Figure 6.4 can be obtained by using identities (6.3) and (6.4), respectively. These observations make it possible to generate the numbers of the spreadsheet of Figure 6.4 from the natural numbers located in row 1 by defining the formula =B1*(B$1+$A2–1)/$A2 in cell B2 and replicating it across rows and columns to cell O9 (see Figure 6.9 in which the spreadsheet of Figure 6.4 is extended to include natural numbers greater than one in column A as referents for the spreadsheet formula).

B2 | fx =B1*(B$1+$A2-1)/$A2

	A	B	C	D	E	F	G	H	I	J	K	L	M	N	O
1		1	2	3	4	5	6	7	8	9	10	11	12	13	14
2	2	1	3	6	10	15	21	28	36	45	55	66	78	91	105
3	3	1	4	10	20	35	56	84	120	165	220	286	364	455	560
4	4	1	5	15	35	70	126	210	330	495	715	1001	1365	1820	2380
5	5	1	6	21	56	126	252	462	792	1287	2002	3003	4368	6188	8568
6	6	1	7	28	84	210	462	924	1716	3003	5005	8008	12376	18564	27132
7	7	1	8	36	120	330	792	1716	3432	6435	11440	19448	31824	50388	77520
8	8	1	9	45	165	495	1287	3003	6435	12870	24310	43758	75582	125970	203490
9	9	1	10	55	220	715	2002	5005	11440	24310	48620	92378	167960	293930	497420

Figure 6.9. Generating binomial coefficients starting from natural numbers.

6.4. A Problem About the Triangle and Its Perimeter

As the first illustration of how external and internal rival factors (Denzin, 1970) can structure the process of triangulation in mathematical problem solving, consider:

Perimeter of Triangle Problem. Perimeter of triangle with three consecutive triangular numbers serving as its side lengths is equal to 136 linear units. Find the side lengths.

Discussion. The problem seeks three consecutive triangular numbers with a given sum. Its algebraic solution requires solving a quadratic equation $\frac{n(n+1)}{2}+\frac{(n+1)(n+2)}{2}+\frac{(n+2)(n+3)}{2}=136$. This can be done with the help of Wolfram Alpha (Figure 6.10) yielding n = 8 as its positive root. Now one can use the spreadsheet of Figure 6.4 to find the three numbers: 36, 45, and 55 (cells H2, I2, and J2, respectively). Alternatively, one can find those numbers in the spreadsheet of Figure 6.6 (cells H6, I7, and J8, respectively).

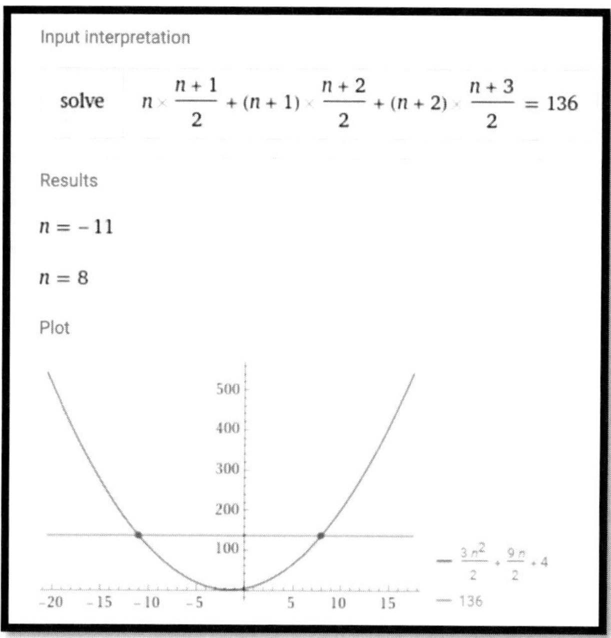

Figure 6.10. Solving a quadratic equation both algebraically and graphically.

The choice of technology to deal with solving an algebraic equation as well as the "ability to make mathematical connections between various approaches to solving problems" (Association of Mathematics Teacher Educators, 2017, p. 31) belong to factors external to the problem. The construction of the above quadratic equation (solved by Wolfram Alpha) requiring skills in algebraization of geometry, the choice of mathematical machinery, and the readiness "to be on the lookout for incomplete or invalid arguments" (Conference Board of the Mathematical Sciences, 2012, p. 33) belong to factors internal to the problem. In connection with the last factor, the method selected to solve the problem is purely symbolic; it does not discuss what makes the problem solvable, that is, what makes the symbol(s) work. In geometric sense, there is no discussion whether such triangle exists; that is, whether the triple of numbers found satisfy the triangle inequality (mentioned in Chapter 1, Section 1.4, and Chapter 3, Section 3.5) – the largest side length is smaller than the sum of the other two side lengths. The above-mentioned triple does satisfy the triangle inequality. Indeed, $55 < 45 + 36$. One can check to see whether any three consecutive triangular numbers satisfy the triangle inequality. A quick look at Figure 6.9 immediately shows that the first two triples (1, 3, 6) and (3, 6, 10) may not be used as the side lengths of a triangle. The graph of Figure 6.11 shows that beginning from the third triple of consecutive triangular numbers, (6, 10, 15), all the triples satisfy the triangle inequality. In other words, the triangle inequality expressed as (using the code of *Wolfram Alpha*)

$$\text{Pochhammer}(k, 2) + \text{Pochhammer}(k + 1, 2) > \text{Pochhammer}(k + 2, 2) \quad (6.5)$$

holds true for $k > \frac{1+\sqrt{17}}{2}$, that is, for any integer $k \geq 3$. At the same time, beginning from the fifth triple of consecutive tetrahedral numbers, (35, 56, 84), all the triples satisfy the triangle inequality (see the 5th tetrahedral number 35 in row 3 of the spreadsheet of Figure 6.4). One can use Wolfram Alpha to further explore triangle inequalities for the triples of numbers from other rows of the spreadsheet of Figure 6.4 by changing the second coordinate of the Pochhammer symbols in inequality (6.5) for 3, 4, 5, and so on. For example, the inequality Pochhammer $(k, 3)$ + Pochhammer$(k + 1, 3) >$ Pochhammer $(k + 2, 3)$ holds true for $k > 1 + \sqrt{10} \cong 4.16$. Figure 6.12 shows the use of Maple in demonstrating that the graph of the function $y = \frac{(x+1)(x+2)(2x+3)}{6}$, representing the sum of two smaller tetrahedral numbers as side lengths, is located above the graph of the function $y = \frac{(x+2)(x+3)(x+4)}{6}$, representing the largest tetrahedral number as a side length, to the right of the point of intersection of the two graphs at the point $x = 1 + \sqrt{10}$. This graphing in Maple confirms what was already found in the contexts of spreadsheet (Figure 6.4) and Wolfram Alpha – beginning from the 5th tetrahedral number, any triple of consecutive tetrahedral numbers may serve as the side lengths of a triangle.

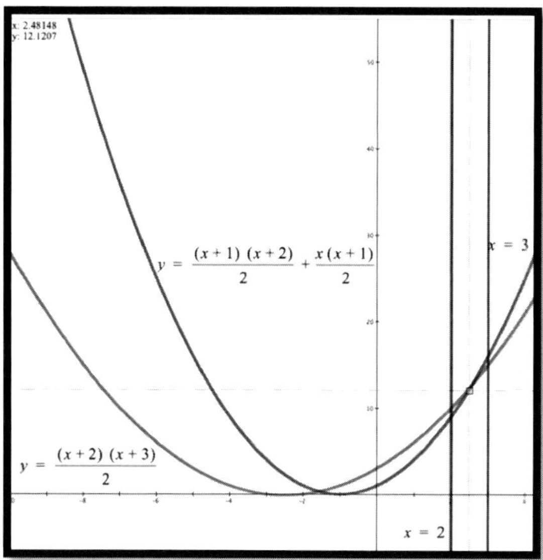

Figure 6.11. Inequality (6.5) in the graphic form using the Graphing Calculator.

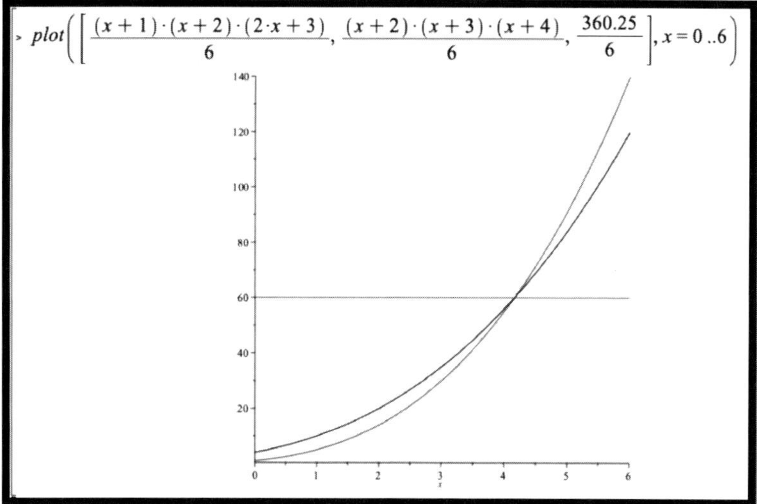

Figure 6.12. Using Maple in triangulating the results of spreadsheet and Wolfram Alpha.

6.5. From Problem Solving to Problem Posing

Changing perimeter of a triangle from 136 to another integer, say, 140 or 132, makes the problem numerically unsolvable when using triangular numbers as the side lengths. This raises another question: How can one find the value of perimeter of a triangle with three consecutive triangular numbers as the side lengths to make the problem solvable? It is through the idea of triangulation in mathematics education as using more than one way of solving a problem and more than one tool in support of problem solving that this question can be answered. As Denzin (1970) put it in the context of sociology, “Problems and questions, not theory, create new perspectives” (p. 55). The same is true for mathematics, in general, and mathematics teacher education, in particular. Questions are the major means of learning mathematics and, according to the National Council of Teachers of Mathematics (2000), “Students’ natural inclination to ask questions must be nurtured ... [even] when the answers are not immediately obvious” (p. 109). As a result, iterative duality of problem solving and posing serves as an agency of triangulation in mathematics education.

Following the recommendation by the Association of Mathematics Teacher Educators (2017) concerned with “unpacking multiple approaches to common mathematical tasks” (p. 90) to which the Perimeter of Triangle Problem belongs, note that it can be solved differently, in a purely visual way, through what may be called geometrization of arithmetic, another internal factor to solving a problem. In order to carry out this method of triangulation, one needs to possess ACU (advanced conceptual understanding, Chapter 3, Section 3.3) of the problem structure. Such understanding can be developed through providing explanation as to why the sum of three consecutive triangular numbers is one greater than a multiple of three. Let t_n be the triangular number of rank n. Then

$$t_n + t_{n+1} + t_{n+2} = \frac{n(n+1) + (n+1)(n+2) + (n+2)(n+3)}{2}$$
$$= \frac{3n^2 + 9n + 8}{2} = 3\frac{n(n+3)}{2} + 1 = 3\left[\frac{n(n+1)}{2} + n + 1\right] + 1$$
$$= 3(t_n + n + 1) + 1 = 3t_{n+1} + 1\,.$$

In particular, the identity

$$3t_{n+1} + 1 = \frac{3n^2+9n+8}{2} \tag{6.6}$$

holds true.

Consider the diagram of Figure 6.13 which demonstrates that adding three times the second triangular number of the triple and increasing it by one yields the perimeter. That is, one can perceive abstract symbol t_n used in the algebraic solution as a concrete (particular) concept embedded into the context of creating a rectangle with a step. This, of course, requires the ability to contextualize by probing into the referents provided by concrete objects. Such ability is also an important aspect of computational thinking – "choosing an appropriate representation for a problem or modeling the relevant aspects of the problem to make it tractable" (Wing, 2006, p. 33). That is, the ideas behind triangulation and computational thinking in mathematical problem solving go hand-in-hand.

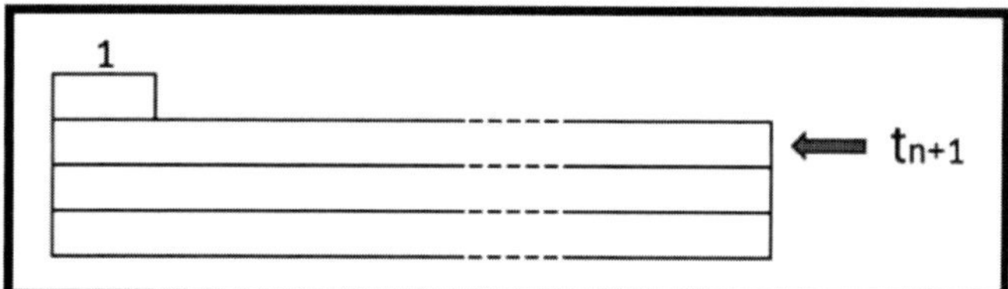

Figure 6.13. The sum of the triangular numbers t_n, t_{n+1}, t_{n+2} is $3t_{n+1} + 1$.

Therefore, to pose a similar problem, one has to select any triangular number greater than or equal to 10 (because, as was mentioned above, the triple (6, 10, 15) is the first one that may be used as side lengths of a triangle), multiply it by three and add one to get the value of perimeter of a triangle. Then, the triple of consecutive triangular numbers the second of which is the selected number would serve as the sought side lengths of the triangle. For example, if one selects the triangular number 55, then $3 \times 55 + 1 = 166$ is the perimeter of the triangle the side lengths of which are 45, 55, and 66.

A similar exploration can be carried out with a triple of consecutive tetrahedral numbers serving as the side lengths of a triangle. To this end, note that $T_{n+1} = T_n + t_{n+1}$. Indeed, in order to get the sum of the first $n + 1$ triangular numbers, T_{n+1}, one has to add t_{n+1} to the sum of the first n triangular numbers. Therefore,

$$T_n + T_{n+1} + T_{n+2} = T_n + T_{n+1} + T_{n+1} + t_{n+2} = T_n + 2T_{n+1} + t_{n+2}$$
$$= T_n + 2(T_n + t_{n+1}) + t_{n+2} = 3T_n + 2t_{n+1} + t_{n+2}.$$

That is, the formula

$$T_n + T_{n+1} + T_{n+2} = 3T_n + 2t_{n+1} + t_{n+2} \tag{6.7}$$

shows that the sum $3T_n + 2t_{n+1} + t_{n+2}$ is the perimeter of a triangle the side lengths of which are tetrahedral numbers T_n, T_{n+1}, T_{n+2}, provided that $n \geq 5$ (see Figure 6.12).

Selecting $T_6 = 56, t_{6+1} = 28, t_{6+2} = 36$, we have $T_6 + T_{6+1} + T_{6+2} = 3 \times 56 + 2 \times 28 + 36 = 260 = 56 + 84 + 120$. That is, if the side lengths of a triangle with perimeter 260 linear units are consecutive tetrahedral numbers, then the side lengths are 56, 84, and 120 linear units.

Furthermore, one can show that

$$3T_n + 2t_{n+1} + t_{n+2} = 3T_{n+1} + n + 2. \tag{6.8}$$

Indeed,

$$3T_n + 2t_{n+1} + t_{n+2} = 3(t_1 + t_2 + \cdots + t_n) + 2t_{n+1} + t_{n+2}$$
$$=2(t_1 + t_2 + \cdots + t_n + t_{n+1}) + t_1 + t_2 + \cdots + t_n + t_{n+1} + n + 2 =$$
$$3T_{n+1} + n + 2\,.$$

For example, when $n = 9$ we have $3T_{9+1} + 9 + 2 = 3 \times 220 + 11 = 671 = 165 + 220 + 286$, implying that a triangle with perimeter 671 linear units may have three consecutive tetrahedral numbers as the side length.

6.6. Computational Solutions to the Perimeter Problem Using Different Tools

One can use a spreadsheet to generate the sums of three consecutive triangular numbers as well as the sums of three consecutive tetrahedral numbers. As was mentioned in the previous section, knowing those numbers makes it possible to pose problems like the Perimeter of Triangle Problem to make such problems solvable. The spreadsheet of Figure 6.14 generates triangular numbers (column B) as partial sums of natural numbers (column A) and then, entering the formula =B1+B2+B3 in cell C1 and replicating it down column C, result in the sequence of the sums of three consecutive triangular numbers (column C). Because the first two triples of triangular numbers may not serve as the side lengths of a triangle (see section 6.3), only beginning from the number 31 (cell C3) all other numbers in column C may be used as perimeters. In particular, one can recognize the number 136 (cell C8) which is the perimeter of the triangle with the side lengths 36, 45, and 55 (discussed in Section 6.4). The spreadsheet of Figure 6.14 generated tetrahedral numbers (as partial sums of triangular numbers) in column E by entering in cell E2 the formula =E1+B2 and replicating it down column E. Consequently, the formula =E1+E2+E3 is entered in cell F1 and replicated down column F, which now contains candidates for the Perimeter of Triangle Problem using tetrahedral numbers as the side lengths. As was mentioned in Section 6.4, only beginning from the fifth triple, (35, 56, 84), that is, beginning from the number 175 (cell F5), the entries of column F may be used as perimeters of triangles with consecutive tetrahedral numbers as the side lengths.

C1 fx =B1+B2+B3

	A	B	C	D	E	F
1	1	1	10		1	15
2	2	3	19		4	34
3	3	6	31		10	65
4	4	10	46		20	111
5	5	15	64		35	175
6	6	21	85		56	260
7	7	28	109		84	369
8	8	36	136		120	505
9	9	45	166		165	671
10	10	55	199		220	870
11	11	66	235		286	1105
12	12	78	274		364	1379
13	13	91	316		455	1695
14	14	105	361		560	2056
15	15	120	409		680	2465
16	16	136	460		816	2925
17	17	153	514		969	3439
18	18	171	571		1140	4010
19	19	190	631		1330	4641
20	20	210	694		1540	5335

Figure 6.14. Finding the sums of three consecutive triangular (column C) and tetrahedral (column F) numbers within a spreadsheet.

In the spirit of computational triangulation, the numbers in column C of the spreadsheet of Figure 6.14 are confirmed by Wolfram Alpha (Figure 6.15) which created a table of integers using the expression $\frac{3n^2+9n+8}{2}$ for $n = 1, 2, 3, \ldots 20$. The same candidates for perimeters of a triangle are generated by Wolfram Alpha (Figure 6.16) using the expression $\frac{3(n+1)(n+2)}{2} + 1$, thereby using triangulation *between* methods (McFee, 1992). Likewise, Wolfram Alpha (Figure 6.17), using the same type of triangulation, confirms the results of the spreadsheet (column F) of Figure 6.14 by using the expression $\frac{3(n+1)(n+2)(n+3)}{6} + n + 2 = \frac{(n+2)(n^2+4n+5)}{2}$ found through combining formulas (6.7) and (6.8).

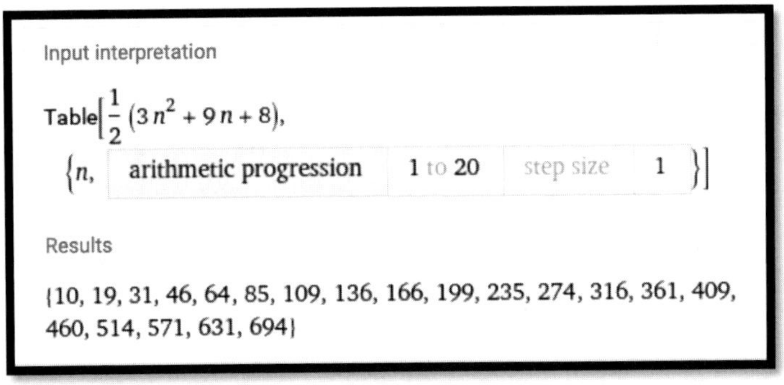

Input interpretation

$\text{Table}\left[\frac{1}{2}(3n^2+9n+8), \{n, \text{ arithmetic progression } 1 \text{ to } 20 \text{ step size } 1\}\right]$

Results

{10, 19, 31, 46, 64, 85, 109, 136, 166, 199, 235, 274, 316, 361, 409, 460, 514, 571, 631, 694}

Figure 6.15. Using *Wolfram Alpha* to find the values of $t_n + t_{n+1} + t_{n+2}$.

Input interpretation

$\text{Table}\left[3(n+1)\times\frac{n+2}{2} + 1, \{n, \text{ arithmetic progression } 1 \text{ to } 20 \text{ step size } 1\}\right]$

Results

{10, 19, 31, 46, 64, 85, 109, 136, 166, 199, 235, 274, 316, 361, 409, 460, 514, 571, 631, 694}

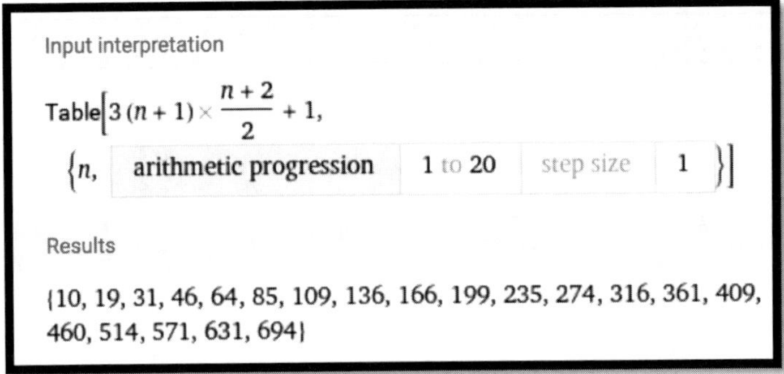

Figure 6.16. Using *Wolfram Alpha* to confirm numerically identity (6.6).

Input interpretation

$$\text{Table}\left[3\,(n+1)\,(n+2)\times\frac{n+3}{6}+n+2,\right.$$
$$\left.\{n,\ \text{arithmetic progression}\ \ 1\ \text{to}\ 20\ \ \text{step size}\ \ 1\}\right]$$

Results

{15, 34, 65, 111, 175, 260, 369, 505, 671, 870, 1105, 1379, 1695, 2056, 2465, 2925, 3439, 4010, 4641, 5335}

Figure 6.17. Using *Wolfram Alpha* to verify (6.8) numerically.

Input interpretation

$$\text{Table}\left[\text{int}\left(\frac{1}{2}\,(3\,n^2+9\,n+8)\right)-10\,\text{int}\left(\frac{1}{20}\,(3\,n^2+9\,n+8)\right),\right.$$
$$\left.\{n,\ \text{arithmetic progression}\ \ 1\ \text{to}\ 200\ \ \text{step size}\ \ 1\}\right]$$

Results

{0, 9, 1, 6, 4, 5, 9, 6, 6, 9, 5, 4, 6, 1, 9, 0, 4, 1, 1, 4, 0, 9, 1, 6, 4, 5, 9, 6, 6, 9, 5, 4, 6, 1, 9, 0, 4, 1, 1, 4, 0, 9, 1, 6, 4, 5, 9, 6, 6, 9, 5, 4, 6, 1, 9, 0, 4, 1, 1, 4, 0, 9, 1, 6, 4, 5, 9, 6, 6, 9, 5, 4, 6, 1, 9, 0, 4, 1, 1, 4, 0, 9, 1, 6, 4, 5, 9, 6, 6, 9, 5, 4, 6, 1, 9, 0, 4, 1, 1, 4, 0, 9, 1, 6, 4, 5, 9, 6, 6, 9, 5, 4, 6, 1, 9, 0, 4, 1, 1, 4, 0, 9, 1, 6, 4, 5, 9, 6, 6, 9, 5, 4, 6, 1, 9, 0, 4, 1, 1, 4, 0, 9, 1, 6, 4, 5, 9, 6, 6, 9, 5, 4, 6, 1, 9, 0, 4, 1, 1, 4, 0, 9, 1, 6, 4, 5, 9, 6, 6, 9, 5, 4, 6, 1, 9, 0, 4, 1, 1, 4, 0, 9, 1, 6, 4, 5, 9, 6, 6, 9, 5, 4, 6, 1, 9, 0, 4, 1, 1, 4}

Figure 6.18. The last digits of the sum $t_n + t_{n+1} + t_{n+2}$ form a 20-cycle.

Finally, using independently Wolfram Alpha (Figure 6.18) and the Graphing Calculator (Figure 6.19) one can demonstrate that the last digits of the candidates for perimeters of triangles with three consecutive triangular numbers as the side lengths form a cycle of length 20. In that case, we have triangulation *within* the method (McFee, 1992) as both tools use the same formula but with different notation (as mentioned in Chapter 3, Section 3.2); namely, in the context of Wolfram Alpha the formula $int\left(\frac{3n^2+9n+8}{2}\right) -$

$10int\left(\frac{3n^2+9n+8}{20}\right)$, and in the context of the Graphing Calculator the formula $floor\left(\frac{3n^2+9n+8}{2}\right) - 10floor\left(\frac{3n^2+9n+8}{20}\right)$. Likewise, using the formulas $int\left(\frac{(n+2)(n^2+4n+5)}{2}\right) - 10int\left(\frac{(n+2)(n^2+4n+5)}{20}\right)$ and $floor\left(\frac{(n+2)(n^2+4n+5)}{2}\right) - 10floor\left(\frac{(n+2)(n^2+4n+5)}{20}\right)$ in the contexts of Wolfram Alpha (Figure 6.20) and the Graphing Calculator (Figure 6.21), respectively, one can demonstrate that the last digits of the candidates for perimeters of triangles with three consecutive tetrahedral numbers as the side lengths form a cycle of length 20 as well. Same results can be confirmed by the spreadsheet in Figure 6.22 and by Maple in Figure 6.23.

A	$B = \frac{3A^2+9A+8}{2}$	$C = \lfloor B \rfloor - 10\left\lfloor \frac{B}{10} \right\rfloor$
1	10	0
2	19	9
3	31	1
4	46	6
5	64	4
6	85	5
7	109	9
8	136	6
9	166	6
10	199	9
11	235	5
12	274	4
13	316	6
14	361	1
15	409	9
16	460	0
17	514	4
18	571	1
19	631	1
20	694	4

Figure 6.19. Another tool shows a 20-cycle formed by the last digits of $t_n + t_{n+1} + t_{n+2}$.

Input interpretation

Table[$\text{int}(0.5\,(n+2)\,(n^2+4n+5)) - 10\,\text{int}(0.05\,(n+2)\,(n^2+4n+5))$,
{n, arithmetic progression 1 to 200 step size 1 }]

Result

{5, 4, 5, 1, 5, 0, 9, 5, 1, 0, 5, 9, 5, 6, 5, 5, 9, 0, 1, 5, 5, 4, 5, 1, 5, 0, 9, 5, 1,
0, 5, 9, 5, 6, 5, 5, 9, 0, 1, 5, 5, 4, 5, 1, 5, 0, 9, 5, 1, 0, 5, 9, 5, 6, 5, 5, 9,
0, 1, 5, 5, 4, 5, 1, 5, 0, 9, 5, 1, 0, 5, 9, 5, 6, 5, 5, 9, 0, 1, 5, 5, 4, 5, 1, 5,
0, 9, 5, 1, 0, 5, 9, 5, 6, 5, 5, 9, 0, 1, 5, 5, 4, 5, 1, 5, 0, 9, 5, 1, 0, 5, 9, 5, 6,
5, 5, 9, 0, 1, 5, 5, 4, 5, 1, 5, 0, 9, 5, 1, 0, 5, 9, 5, 6, 5, 5, 9, 0, 1, 5, 5, 4, 5,
1, 5, 0, 9, 5, 1, 0, 5, 9, 5, 6, 5, 5, 9, 0, 1, 5, 5, 4, 5, 1, 5, 0, 9, 5, 1, 0, 5, 9,
5, 6, 5, 5, 9, 0, 1, 5, 5, 4, 5, 1, 5, 0, 9, 5, 1, 0, 5, 9, 5, 6, 5, 5, 9, 0, 1, 5}

Figure 6.20. The last digits of the sum $T_n + T_{n+1} + T_{n+2}$ form a 20-cycle.

A	$B = (A+2)\dfrac{(A^2+4A+5)}{2}$	$C = \lfloor B \rfloor - 10\left\lfloor \dfrac{B}{10} \right\rfloor$
1	15	5
2	34	4
3	65	5
4	111	1
5	175	5
6	260	0
7	369	9
8	505	5
9	671	1
10	870	0
11	1105	5
12	1379	9
13	1695	5
14	2056	6
15	2465	5
16	2925	5
17	3439	9
18	4010	0
19	4641	1
20	5335	5

Figure 6.21. Another tool shows a 20-cycle formed by the last digits of $T_n + T_{n+1} + T_{n+2}$.

	A	B	C	D	E	F	G	H	I	J	K	L	M	N	O	P	Q	R	S	T
1	1	2	3	4	5	6	7	8	9	10	11	12	13	14	15	16	17	18	19	20
2																				
3	0	9	1	6	4	5	9	6	6	9	5	4	6	1	9	0	4	1	1	4
4																				
5	5	0	3	7	5	0	5	3	7	0	5	5	3	2	5	5	5	8	7	5

Figure 6.22. Using a spreadsheet to confirm the results shown in Figures 6.17 – 6.20.

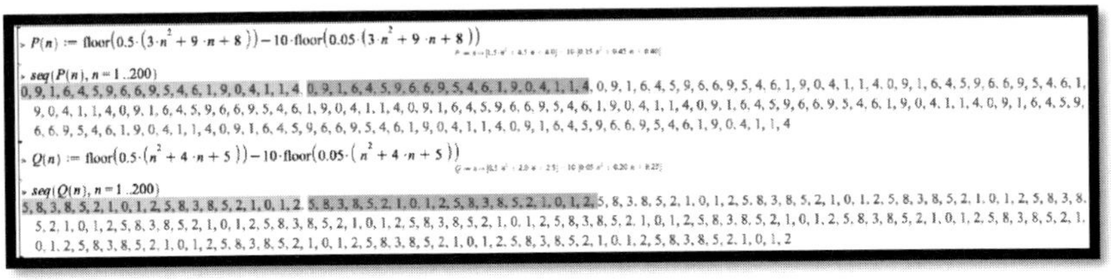

```
> P(n) := floor(0.5·(3·n^2 + 9·n + 8)) − 10·floor(0.05·(3·n^2 + 9·n + 8))
> seq(P(n), n = 1..200)
0, 9, 1, 6, 4, 5, 9, 6, 6, 9, 5, 4, 6, 1, 9, 0, 4, 1, 1, 4, 0, 9, 1, 6, 4, 5, 9, 6, 6, 9, 5, 4, 6, 1, 9, 0, 4, 1, 1, 4, 0, 9, 1, 6, 4, 5, 9, 6, 6, 9, 5, 4, 6, 1, 9, 0, 4, 1, 1, 4, 0, 9, 1, 6, 4, 5, 9, 6, 6, 9, 5, 4, 6, 1,
9, 0, 4, 1, 1, 4, 0, 9, 1, 6, 4, 5, 9, 6, 6, 9, 5, 4, 6, 1, 9, 0, 4, 1, 1, 4, 0, 9, 1, 6, 4, 5, 9, 6, 6, 9, 5, 4, 6, 1, 9, 0, 4, 1, 1, 4, 0, 9, 1, 6, 4, 5, 9, 6, 6, 9, 5, 4, 6, 1, 9, 0, 4, 1, 1, 4, 0, 9, 1, 6, 4, 5, 9,
6, 6, 9, 5, 4, 6, 1, 9, 0, 4, 1, 1, 4, 0, 9, 1, 6, 4, 5, 9, 6, 6, 9, 5, 4, 6, 1, 9, 0, 4, 1, 1, 4, 0, 9, 1, 6, 4, 5, 9, 6, 6, 9, 5, 4, 6, 1, 9, 0, 4, 1, 1, 4
> Q(n) := floor(0.5·(n^2 + 4·n + 5)) − 10·floor(0.05·(n^2 + 4·n + 5))
> seq(Q(n), n = 1..200)
5, 8, 3, 8, 5, 2, 1, 0, 1, 2, 5, 8, 3, 8, 5, 2, 1, 0, 1, 2, 5, 8, 3, 8, 5, 2, 1, 0, 1, 2, 5, 8, 3, 8, 5, 2, 1, 0, 1, 2, 5, 8, 3, 8, 5, 2, 1, 0, 1, 2, 5, 8, 3, 8, 5, 2, 1, 0, 1, 2, 5, 8, 3, 8, 5, 2, 1, 0, 1, 2, 5, 8, 3, 8,
5, 2, 1, 0, 1, 2, 5, 8, 3, 8, 5, 2, 1, 0, 1, 2, 5, 8, 3, 8, 5, 2, 1, 0, 1, 2, 5, 8, 3, 8, 5, 2, 1, 0, 1, 2, 5, 8, 3, 8, 5, 2, 1, 0, 1, 2, 5, 8, 3, 8, 5, 2, 1, 0, 1, 2, 5, 8, 3, 8, 5, 2, 1, 0, 1, 2, 5, 8, 3, 8, 5, 2, 1,
0, 1, 2, 5, 8, 3, 8, 5, 2, 1, 0, 1, 2, 5, 8, 3, 8, 5, 2, 1, 0, 1, 2, 5, 8, 3, 8, 5, 2, 1, 0, 1, 2, 5, 8, 3, 8, 5, 2, 1, 0, 1, 2, 5, 8, 3, 8, 5, 2, 1, 0, 1, 2
```

Figure 6.23. Using Maple to confirm the results shown in Figures 6.17 – 6.21.

Conclusion

The chapter was motivated by Bowen's study of emotional systems using the concept of triangulation. This allowed for mathematical investigations through mathematising real-life situations involving more than four individuals. The sequences of natural, triangular, and tetrahedral numbers, and connections among them were discussed. Using a problem about triangle and its perimeter, conditions allowing for three consecutive terms of such sequences to serve as the side lengths that satisfy the triangle inequality were developed to enable natural transition from problem solving to problem posing. The scope of this chapter is appropriate to be included in a course for prospective secondary mathematics teachers. The next chapter will use computational triangulation and the TITE framework in exploring various subsequences of natural numbers called number sieves.

Chapter 7

Computational Triangulation in Elementary Number Theory

7.1. Introduction

Representation of numbers as sums of other numbers is one of the big ideas of mathematics. At the very basic level, an even number is a sum of two odd numbers. Also, a square number of rank n is the sum of the first n odd numbers, e.g., $4^2 = 1 + 3 + 5 + 9$, something that at the pre-school age was noticed by a Russian mathematician Andrey Nikolaevich Kolmogorov (1903-1987), one of the major contributors to mathematics of the 20th century, who considered this observation as his first mathematical discovery (Tikhomirov, 2001). Some squares are the sums of two other squares (the Pythagorean theorem). A more complicated case is the Goldbach's conjecture, formulated by a German mathematician Christian Goldbach (1690-1764), in a letter to Euler, stating that every even number greater than 2 is a sum of two prime numbers, remains (by the time of writing this book) an unsolved problem of the elementary theory of numbers. The Tertiary Goldbach conjecture stating that every odd number greater than five is the sum of three prime numbers (e.g., 15 = 3 + 5 + 7), that, according to Vavilov (2021), was first formulated by a French mathematician and philosopher René Descartes (1596-1650), was proved only recently by Helfgott (2014). That is, the representation of numbers as sums of other numbers is both classic and contemporary domain of mathematical research. Many mathematics education problems deal with such representations and are often discussed through the lens of computation. Among them are expectations for kindergarten students to "decompose numbers less than 10 into pairs" (Common Core State Standards, 2010, p. 11), for 2nd graders to "write an equation to express an even number as a sum of two equal addends" (ibid, p. 19), for 4th graders "decomposing fractions into unit fractions" (ibid, p. 27), and "when many or cumbersome computations are needed to solve problems … calculator should be set aside to allow this focus" (National Council of Teachers of Mathematics, 2000, pp. 32, 33). An elementary teacher candidate (the author's non-traditional student) connected these standards-based expectations for the learners of mathematics to knowing history of the subject matter by recognizing the changing practice of teaching and learning mathematics over the years. As the candidate put it, "*students' needs are always changing in education and the way we teach should be changing also. We need to grow with the students with the different ways they learn. The history of math is important for every teacher to know. Just because they know how to do math doesn't mean they know the proper way to learn it.*" This comment from the field confirms the need of integrating history of mathematics in teacher education courses in the form of "historical snippets" (Tzanakis & Arcavi, 2002, p. 214), thereby presenting "mathematics as a living and evolving subject" (Conference Board of the Mathematical Sciences, 2012, p. 61). As noted by mathematics educators in Chile, focusing on historical developments of mathematical ideas provides teacher candidates with information and cultural awareness "that facilitate understanding of the subject matter and that may be used as a teaching material" (Felmer et al., 2014, p. 36). Furthermore, in the spirit of Klein's (2016) description of mathematics for teachers from a higher standpoint, paying attention to "historical development of mathematical concepts … allows prospective teachers to "stand above" their subject" (Bernardi & Menghini, 2024, p. 34)

7.2. From Arithmetic Sequences to Polygonal Numbers to Number Sieves

Polygonal numbers, of which a square is a special case, follow the suit. Polygonal numbers can be defined as partial sums of arithmetic sequences. Indeed, the left-hand side of the equality $1 + 3 + 5 + 9 = 16$, defining the fourth square number, is a partial sum of an arithmetic sequence with the first term one and difference two (i.e., the sequence of odd numbers). Likewise, the left-hand side of the equality $1 + 2 + 3 + 4 = 10$, defining the fourth triangular number, is a partial sum of an arithmetic sequence with the first term one and difference one (i.e., the sequence of natural numbers). That is, the difference of an arithmetic sequence defining an m-gonal number is $m - 2$. The n-th partial sum of an arithmetic sequence $a_n = a_1 + d(n-1)$ with the first term $a_1 = 1$ and difference $d = m - 2$, defining $P(m, n)$ – the m-gonal number of rank n, can be found as the sum of the arithmetic sequence $\frac{a_1+a_n}{2}n = \frac{2a_1+(m-2)(n-1)}{2}n = \frac{2+(m-2)(n-1)}{2}n = \frac{(m-2)(n-1)n}{2} + n$. For example, $\left.\frac{(m-2)(n-1)n}{2}\right|_{m=4} + n = (n-1)n + n = n^2$. That is,

$$P(m,n) = \frac{(m-2)(n-1)n}{2} + n\,, m \geq 3, n \geq 1. \quad (7.1)$$

Note that tetrahedral numbers were defined as partial sums of triangular numbers which, however, do not form an arithmetic sequence as the difference between two consecutive triangular numbers is not a constant but the rank of the larger number. Whereas polygonal numbers are quadratic functions of their ranks, tetrahedral numbers are cubic functions of their ranks (see formula (1.2), Chapter 1, Section 1.4).

The term polygonal number is probably best known in mathematics from the conjecture by a French mathematician Pierre de Fermat (1607-1665) that every natural number is a sum of at most m polygonal numbers of side m. Such representations are not necessarily unique. For example, whereas 21 = 16 + 4 + 1 only, 22 = 16 + 4 + 1 + 1 = 9 + 9 + 4, where the numbers 1, 4, 9 and 16 are square numbers or polygonal numbers of side four and ranks 1, 2, 3, and 4, respectively. Also, whereas 31 = 12 + 12 + 5 + 1 + 1 only, 41 = 22 + 12 + 5 + 1 + 1 = 12 + 12 + 12 + 5 = 35 + 5 + 1, where the numbers 1, 5, 12, 22, and 35 are pentagonal numbers or polygonal numbers of side five and ranks 1, 2, 3, 4, and 5, respectively. That is, any natural number can be represented through a sum of four or fewer square numbers and five or fewer pentagonal numbers, perhaps in more than one way.

When m = 3 we have triangular numbers, known from the time of Pythagoras (570 B.C.-490 B.C.) as partial sums of consecutive natural numbers (Beiler, 1966). The term triangular number is familiar to mathematicians from the theorem proved in 1796 by Carl Friedrich Gauss (1777-1855, Germany), commonly regarded being the greatest mathematician of all time, that any natural number is a sum of at most three triangular numbers (Andrews, 1986). The general case, known as the polygonal number theorem, was proved by a notable French mathematician Augustin-Louis Cauchy (1789-1857) some fifteen years later (Nathanson, 1987). In other words, triangular and, in general, m-gonal numbers form an additive basis of orders three and m, respectively, for natural numbers (Bell et al., 2018). In mathematics education, triangular numbers connect arithmetic to geometry, as such sums can be represented in the form of evolving equilateral triangles representing a special case of the partial sums of arithmetic sequences. Principles and Standards for School Mathematics introduce triangular numbers as a context that "illustrates what reasoning and proof can look like in the middle grades" (National Council of Teachers of Mathematics, 2000, p. 264).

According to Gauss (1966), "The problem of distinguishing prime numbers from composite numbers and of resolving the latter into their prime factors is known to be one of the most important and useful in arithmetic" (p. 396). From ancient times, prime numbers are known to be separated from composite numbers by applying the so-called sieve process which bears the name of a Greek scholar Eratosthenes (c.276 B.C-c.195 B.C.). The sieve of Eratosthenes is based on the idea of removing, step-by-step, multiples of primes (i.e., integers with more than two divisors) from natural numbers (greater than one). The word sieve to be used in this chapter is borrowed from this context and a separation process can be applied to polygonal numbers as well. For example, removing any other number (the rule of the sieve process in that case) from the sequence of triangular numbers 1, 3, 6, 10, 15, 21, 28, ... – polygonal numbers of side three yields the sequence of hexagonal numbers 1, 6, 15, 28, ... – polygonal numbers of side six, that are also triangular numbers of odd ranks. Formulating mathematically the ranks of the remaining terms of a sequence to which the sieve is applied, one can easily computerize the process using a spreadsheet, one of the tools used for informal demonstration of mathematical propositions.

Informal demonstration can be traced back in the history of mathematics to antient Greece. It is known that Archimedes (287 B.C-212 B.C.), by some accounts (Rohlin, 2013) one of the greatest mathematicians of all time, wrote to Eratosthenes: "Certain things first became clear to me by a mechanical method, although they had to be demonstrated by geometry afterwards because their investigation by the said mechanical method did not furnish an actual demonstration. But it is of course easier, when the method has previously given us some knowledge of the questions, to supply the proof than it is to find it without any previous knowledge" (Archimedes, 1912, p. 13). Nowadays, what Archimedes called a mechanical method is the use of digital tools capable of initiating the interplay between computations and mathematics. This interplay is in the duality of mathematical and computational methods in a sense that whereas computations facilitate access to mathematical knowledge, mathematics itself is used to improve the efficiency of computations, which, in turn, enable advancements in various applications of mathematics, including education, engineering and science. In the modern-day discourse, the reference to geometry by Archimedes as a rigorous demonstration (alternatively, Pólya's (1954) reference to "formal proof" (p. 51) – see Chapter 1, Section 1.2) can be substituted by computational triangulation supported by more than one digital instrument capable of symbolic computations, replacing tedious paper and pencil algebraic transformation, and more than one problem-solving approach that structure the algebraic symbolism of computations. In the context of triangular (or, more generally, polygonal) numbers, the activity of developing sieves of such numbers means crossing out every other term from the original number sequence, then every other term from the so developed subsequence, and so on; thus, arriving on the k-th step of this elimination process to the sieve of order k. Other, more complicated types of sieves will be considered also in the last section of this chapter.

7.3. Describing Computational Problem-Solving Strategies Used in This Chapter

In what follows, four problem-solving strategies used to mathematise computationally inquiries into the sieves of triangular numbers will be presented. The first strategy is to use Wolfram Alpha in making a transition from the first few terms of a number sequence developed at each step of a sieve process to its possible numeric continuation and symbolic identification. The next step within this strategy is to generate closed-form formulas for numeric sequences in order to have a single algebraic formula for the entire family of numeric sequences called the triangular number sieve of order k obtained through k, step by step, sieve processes. The process of generalization will be based on the TITE framework introduced in Chapter 3, Section 3.5, as a combination of digital computations and epistemic-oriented interpretation of those pragmatically obtained computational results.

Such combination of the tool use and its epistemology allows for Wolfram Alpha, being an element of the instrumental act (Vygotsky, 1930), to be inserted between a problem solver (a subject) and a problem (an object) and act bi-directionally (see Chapter 1, Section 1.3). In one direction, the tool acts primarily toward solving a problem through a digital computation. At the same time, the tool's utilization is not always straightforward for it requires various intellectual efforts on the part of the subject, thereby, initiating another direction toward one's cognitive development. For example, if the problem is to find the sum of the first 99 odd numbers, one can enter in the input box of Wolfram Alpha the command "sum of the first 99 odd numbers" to get 9801 as the answer. At the same time, one can recognize in the problem an epistemic connection to square numbers (see the beginning of this chapter) and use the command "99^2" which yields 9801 as well. That is, whereas the tool's role is computing, another possible role of the tool is to stimulate mental processes associated with the problem that have potential to fundamentally change the course of using the tool.

The second strategy to be used in the development of an algebraic formula describing the triangular number sieve of order *k* consists of the study of ranks of triangular numbers that constitute the specific sieve through connecting the original ranks with those provided by the sieve. Through this strategy, one can develop an alternative form of a second-degree polynomial representing the triangular number sieve of order *k* and then use another digital tool – Maple – for proving computationally that different symbolic formulas developed for the sieve are identical in a sense that they generate same integers as were provided by Wolfram Alpha. This analytically driven computational triangulation provides rigor for the development of symbolic representations of triangular number sieves. Already in the context of preparing to teach mathematics at the middle school level, teacher candidates "must have opportunities to engage in the use of a variety of technological tools ... to explore and deepen their understanding of mathematics, even if those tools are not the same ones they will eventually use with students" (Conference Board of the Mathematical Sciences, 2012, p. 50).

The third problem-solving strategy is to use a spreadsheet in generating triangular number sieves of different orders and using computational triangulation in verifying the accuracy of symbolic computations provided by Wolfram Alpha and Maple. This strategy enables another generalization and the development of a closed-form formula for a polygonal number sieve of order *k* which includes triangular, square, pentagonal, hexagonal, and other sieves of polygonal numbers.

The fourth problem-solving strategy deals with a mathematical interpretation of numerical modeling of polygonal number sieves that, by recourse to the OEIS®, a rich source of information about integer sequences, leads to the discovery of a special type of sequence. This new sequence represents a new polygonal number sieve the terms of which grow fast due to an exponential growth of numbers to be eliminated at each step. Commonalities among new sieves developed from polygonal numbers of different sides are explained in the context of the fourth strategy through computational triangulation.

7.4. The First Problem-Solving Strategy of Constructing Triangular Number Sieves

As was mentioned in section 7.2, triangular numbers can be defined as the partial sums of consecutive natural numbers 1, 1 + 2 = 3, 1 + 2 + 3 = 6, 1 + 2 + 3 + 4 = 10, and so on. In that way, the sequence of triangular numbers 1, 3, 6, 10, 15, 21, ... develops. In order to find a closed-form formula for triangular numbers, one can enter their first four terms into the input box of Wolfram Alpha to get $a_n = \frac{n(n+1)}{2}$ (Figure 7.1) which, for consistency with the notation that follows, can be written as

$$t_{n,0} = \frac{1}{2}n^2 + \frac{1}{2}n. \quad (7.2)$$

As it was also mentioned in section 7.2, every other triangular number is a hexagonal number

$$1, 6, 15, 28, 45, \ldots \quad (7.3)$$

The closed-form formula for hexagonal numbers generated by Wolfram Alpha (Figure 7.2) is

$$t_{n,1} = 2n^2 - n. \quad (7.4)$$

The sequence of hexagonal numbers (7.3) can be called the triangular number sieve of order one obtained by the rule of crossing out from the sequence 1, 3, 6, 10, 15, 21, 28, ... every other number to get the sequence 1, 6, 15, 28, ... In order to develop the triangular number sieve of order two, one has to cross out every other number from sequence (7.3) to get the sequence

$$1, 15, 45, 91, 153, \ldots . \quad (7.5)$$

To find a closed-form formula for sequence (7.5), one can enter its first four terms into the input box of Wolfram Alpha to get (Figure 7.3)

$$t_{n,2} = 8n^2 - 10n + 3. \quad (7.6)$$

Input interpretation

$\{1, 3, 6, 10, \ldots\}$

Possible sequence identification

Closed form

$a_n = \frac{1}{2} n(n+1)$ (for all terms given)

Continuation

1, 3, 6, 10, 15, 21, 28, 36, 45, 55, 66, 78, 91, 105, 120, 136, 153, 171, ...

Figure 7.1. Finding a closed-form formula for the triangular number sieve of order zero.

Input interpretation

$\{1, 6, 15, 28, \ldots\}$

Possible sequence identification

Closed form

$a_n = 2n^2 - n$ (for all terms given)

Continuation

1, 6, 15, 28, 45, 66, 91, 120, 153, 190, 231, 276, 325, 378, 435, 496, ...

Figure 7.2. Finding a closed-form formula for the triangular number sieve of order one.

Input interpretation

$\{1, 15, 45, 91, \ldots\}$

Possible sequence identification

Closed form

$a_n = 8n^2 - 10n + 3$ (for all terms given)

Continuation

1, 15, 45, 91, 153, 231, 325, 435, 561, 703, 861, 1035, 1225, 1431, ...

Figure 7.3. Finding a closed-form formula for the triangular number sieve of order two.

Continuation of the sequence defined by formula (7.6) will be verified as a special case over the corresponding general sequence generated by Wolfram Alpha through the second problem-solving strategy (Figure 7.10, the third column of the matrix Result) and by a spreadsheet through the third problem-solving strategy (Figure 7.12, column E). In turn, crossing out every other number from sequence (7.5) yields the triangular number sieve of order three

$$1, 45, 153, 325, 561, \dots, \tag{7.7}$$

the closed-form formula of which can be generated by Wolfram Alpha (Figure 7.4) in the form

$$t_{n,3} = 32n^2 - 52n + 21. \tag{7.8}$$

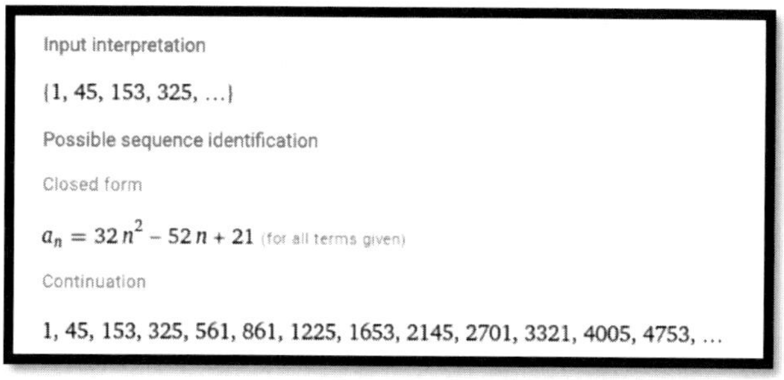

Input interpretation

{1, 45, 153, 325, ...}

Possible sequence identification

Closed form

$a_n = 32 n^2 - 52 n + 21$ (for all terms given)

Continuation

1, 45, 153, 325, 561, 861, 1225, 1653, 2145, 2701, 3321, 4005, 4753, ...

Figure 7.4. Finding a closed-form formula for the triangular number sieve of order three.

Continuation of the sequence defined by formula (7.8) will be verified as a special case over the corresponding general sequence generated by Wolfram Alpha through the second problem-solving strategy (Figure 7.10, the fourth column of the matrix Result) and by a spreadsheet through the third problem-solving strategy (Figure 7.12, column F). Crossing out every other number from sequence (7.7) yields the triangular number sieve of order four

$$1, 153, 561, 1225, 2145, \dots \tag{7.9}$$

with the closed-form formula (generated by Wolfram Alpha as shown in Figure 7.5)

$$t_{n,4} = 128n^2 - 232n + 105. \tag{7.10}$$

Input interpretation

{1, 153, 561, 1225, ...}

Possible sequence identification

Closed form

$a_n = 128n^2 - 232n + 105$ (for all terms given)

Continuation

1, 153, 561, 1225, 2145, 3321, 4753, 6441, 8385, 10585, 13041, ...

Figure 7.5. Finding a closed-form formula for the triangular number sieve of order four.

Continuation of the sequence defined by formula (7.10) will be verified as a special case over the corresponding general sequence generated by Wolfram Alpha through the second problem-solving strategy (Figure 7.10, the fifth column of the matrix Result) and by a spreadsheet through the third problem-solving strategy (Figure 7.12, column G). Likewise, the formula

$$t_{n,5} = 512n^2 - 976n + 465 \tag{7.11}$$

can be developed by Wolfram Alpha for the triangular number sieve of order five (Figure 7.6). Continuation of the sequence defined by formula (7.11) will be verified as a special case over the corresponding general sequence generated by Wolfram Alpha through the second problem-solving strategy (Figure 7.10, the sixth column of the matrix Result) and by a spreadsheet through the third problem-solving strategy (Figure 7.12, column H).

Input interpretation

{1, 561, 2145, 4753, ...}

Possible sequence identification

Closed form

$a_n = 512n^2 - 976n + 465$ (for all terms given)

Continuation

1, 561, 2145, 4753, 8385, 13041, 18721, 25425, 33153, 41905, ...

Figure 7.6. Finding a closed-form formula for the triangular sieve of order five.

Observing formulas (7.2), (7.4), (7.6), (7.8), (7.10) and (7.11), one can note that the coefficients in n^2 are $2^{-1}, 2^1, 2^3, 2^5, 2^7, 2^9$. The sequence of exponents -1, 1, 3, 5, 7, 9, ... can be written in the form $a_k = 2k - 1$, where k = 0, 1, 2, 3, ... are the values of the order of a sieve. The sequence of the coefficients in n^1 are 1/2, -1, -10, -52, -976, ... Using Wolfram Alpha (Figure 7.7), one gets the formula $b_k = 2^{2k} - 3 \times 2^{k-1}$. The sequence of the coefficients in n^0 are 0, 0, 3, 21, 105, 465, ... Using Wolfram Alpha (Figure 7.8), one gets the formula

$$c_k = \frac{2^{2k} - 3 \times 2^{k+1} + 8}{8} = \frac{(2^k - 4) \times (2^k - 2)}{8}.$$

Input interpretation

$\text{Table}\left[2^{2k} - 3 \times 2^{k-1}, \{k, \text{ arithmetic progression } 0 \text{ to } 20 \text{ step size } 1\}\right]$

Result

$\left\{-\frac{1}{2}, 1, 10, 52, 232, 976, 4000, 16192, 65152, 261376, 1047040, 4191232, 16771072, 67096576, 268410880, 1073692672, 4294868992, 17179672576, 68719083520, 274877120512, 1099510054912\right\}$

Figure 7.7. General form of coefficients in n^1 of the triangular number sieve of order k.

Input interpretation

{0, 0, 3, 21, 105, 465, ...}

Possible sequence identification

Closed form

$a_n = \frac{1}{8}\left(2^{2n} - 3\ 2^{n+1} + 8\right)$ (for all terms given)

Continuation

0, 0, 3, 21, 105, 465, 1953, 8001, 32385, 130305, 522753, 2094081, ...

Figure 7.8. General form of coefficients in n^0 of the triangular number sieve of order k.

These computations prompt generalization to have a closed-form formula for the triangular number sieve of order k. Using Wolfram Alpha (Figure 7.10), generalization yields the formula

$$t_{n,k} = 2^{2k-1}n^2 - (2^{2k} - 3 \times 2^{k-1})n + (2^k - 1)(2^{k-1} - 1), n \geq 1, k \geq 0 . \tag{7.12}$$

Remark 7.1. One can use the OEIS® to get closed-form formulas for triangular number sieves of orders one, two, three, four, by entering the first few terms of sequences (7.6), (7.8), (7.10), (7.11) into the input box of the OEIS®. It turns out that sequence (7.6) is included into OEIS®, yet the tool offers the quadratics $8n^2 + 6n + 1$, rather than $8n^2 - 10n + 3$ defined by relation (7.7). Nonetheless, the former turns into the latter after substituting (in the former) $n - 1$ for n. Indeed, $8(n-1)^2 + 6(n-1) + 1 = 8n^2 - 10n + 3$. That is, $(8n^2 + 6n + 1)|_{n=0} = (8n^2 - 10n + 3)|_{n=1} = 1$. At the same time, whereas triangular number sieves defined numerically by (7.7), and symbolically by (7.8), are parts of the OEIS®, it offers their closed-form identical to what is offered by Wolfram Alpha.

Remark 7.2. One can use Maple (Figure 7.9) to replace closed-form formula (7.12) by recursive formula for the triangular number sieve of order k. This formula and its initial condition are

$$t_{n+1,k} = t_{n,k} + \frac{4^k(2n-1)+3\cdot 2^k}{2}, \; t_{0,k} = (2^k - 1)(2^{k-1} - 1). \tag{7.13}$$

In particular, when $k = 0$ we have $t_{n+1,0} = t_{n,0} + (n+1)$ and when $n = k = 0$ we have $t_{1,0} = t_{0,0} + 1 = 1$.

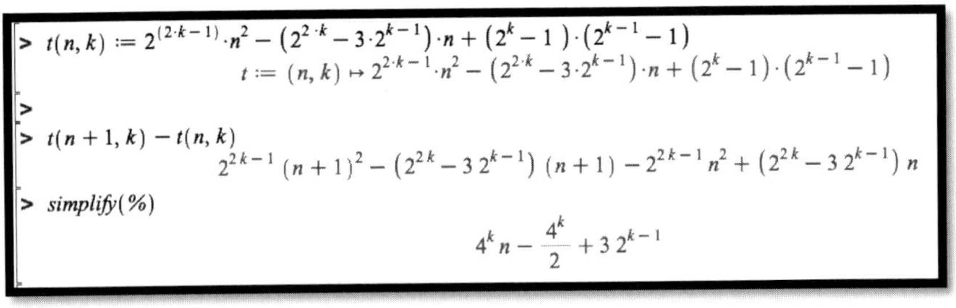

> $t(n,k) := 2^{(2\cdot k-1)}\cdot n^2 - (2^{2\cdot k} - 3\cdot 2^{k-1})\cdot n + (2^k - 1)\cdot(2^{k-1} - 1)$

$$t := (n,k) \mapsto 2^{2\cdot k-1}\cdot n^2 - (2^{2\cdot k} - 3\cdot 2^{k-1})\cdot n + (2^k - 1)\cdot(2^{k-1} - 1)$$

>

> $t(n+1,k) - t(n,k)$

$$2^{2k-1}(n+1)^2 - (2^{2k} - 3\,2^{k-1})(n+1) - 2^{2k-1}n^2 + (2^{2k} - 3\,2^{k-1})\,n$$

> *simplify*(%)

$$4^k n - \frac{4^k}{2} + 3\,2^{k-1}$$

Figure 7.9. Using *Maple* in developing recursive formula (7.13).

Remark 7.3. Formula (7.1) shows that a polygonal number is a quadratic function of its rank and, therefore, the first four numbers of a triangular sieve (starting from the sieve of order zero) fully define the corresponding quadratic formula. Unlike polygonal numbers, Fibonacci numbers, according to formula (3.4), depend on their ranks through the exponents of the roots of a quadratic equation. That is why, as was mentioned in Chapter 3, Section 3.5, entering only the four Fibonacci numbers (2, 3, 5, 8) into the input box of Wolfram Alpha resulted in recognition of a quadratic sequence rather than Fibonacci numbers. One can note that the success of using limited numeric evidence in the case of triangular number sieves, as demonstrated by the corresponding formulas, was due to ACU that enabled the recognition of this difference between polygonal and Fibonacci numbers by a digital tool. As an aside, note that one can use Wolfram Alpha to see that the four numbers 1, 6, 10, 28 – the triangular numbers of ranks 1, 3, 4 and 7 (which are neither consecutive natural numbers nor integers in arithmetic progression) – do not define a quadratic formula, unlike 1, 6, 15, 28 – the first four elements of the triangular sieve of order one which, as shown by formula (7.4), define a quadratic sequence.

Remark 7.4. Although a quadratic function $f(n) = an^2 + bn + c$ is fully defined by three values of $f(n)$, Wolfram Alpha needs four consecutive terms of a triangular number sieve to determine the values of a, b, and c. Alternatively, knowing the values of $f(1), f(2),$ and $f(3)$, one can use Wolfram Alpha to solve the system of the equations (using the command "solve the system of 3 equations") $a + b + c = f(1), 4a + 2b + c = f(2), 9a + 3b + c = f(3)$, to find $f(n)$. This applies to any three values of $f(n)$. For example, the (non-consecutive) triangular numbers 1, 6, 28, define the quadratic sequence $f(n) = \frac{17}{2}n^2 - \frac{41}{2}n + 13$, which does not represent a triangular number sieve as defined at the conclusion of section 7.2.

7.5. The Second Problem-Solving Strategy of Constructing Triangular Number Sieves

The question to be explored in this section is: Where did exponents with the base two appearing in formulas (7.12) and (7.13) come from? Of course, whereas the elements of the triangular number sieves as triangular numbers are the second powers of their ranks within a sieve, it is due to the coefficients of the sieve that the growth of its elements takes place. Another approach to developing a closed-form formula for the elements of the triangular number sieve of order k is to look for the ranks of triangular numbers which form the sieves of order k, $k = 0, 1, 2, 3, \ldots$.

The ranks of the elements of the triangular number sieve of order zero (that is, of the sequence of triangular numbers) are 1, 2, 3, 4, 5, ... and can be defined as

$$r_{n,0} = 2^0(n-1)+1, n = 1,2,3,\ldots . \tag{7.14}$$

For example, when $n = 5$, formula (7.14) yields $r_{5,0} = 2^0(5-1)+1 = 5$ – the rank of the triangular number 15.

The ranks of the elements of the triangular number sieve of order one (that is, the ranks of the hexagonal numbers 1, 6, 15, 28, 45, ... are 1, 3, 5, 7, 9, ...) can be defined as

$$r_{n,1} = 2^1(n-1)+1, n = 1,2,3,\ldots . \tag{7.15}$$

For example, when $n = 5$, formula (7.15) yields $r_{5,1} = 2^1(5-1)+1 = 9$ – the rank of the triangular number 45.

The ranks of the elements of the triangular number sieve of order two (or hexagonal number sieve of order one; that is, the ranks of the triangular numbers 1, 15, 45, 91, 153, ... are 1, 5, 9, 13, 17, ...) can be defined as

$$r_{n,2} = 2^2(n-1)+1, n = 1,2,3,\ldots . \tag{7.16}$$

For example, when $n = 5$, formula (7.16) yields $r_{5,2} = 2^2(5-1)+1 = 17$ – the rank of the triangular number 153.

The ranks of the elements of the triangular number sieve of order three (or hexagonal number sieve of order two; that is, the ranks of the triangular numbers 1, 45, 153, 325, 561, … are 1, 9, 17, 25, 33, …) can be defined as

$$r_{n,3} = 2^3(n-1)+1, n = 1,2,3,\ldots . \tag{7.17}$$

For example, when $n = 5$, formula (7.17) yields $r_{5,3} = 2^3(5-1)+1 = 33$ – the rank of the triangular number 561.

Generalizing from formulas (7.14) – (7.17) to the ranks of the elements of the triangular number sieve of order k yields the sequence of the ranks

$$r_{n,k} = 2^k(n-1)+1, n = 1,2,3,\ldots, k = 0,1,2,\ldots . \tag{7.18}$$

Therefore, by combining formulas (7.18) and (7.2), the n-th element of the triangular number sieve of order k can be written in the form

$$t_{n,k} = \frac{1}{2}\,[2^k(n-1)+1][2^k(n-1)+2]\ ,\ n = 1,2,3,\ldots, k = 0,1,2,\ldots \tag{7.19}$$

For example, when $n = 5$ and $k = 3$, formula (7.19) yields

$t_{5,3} = \frac{1}{2}\,[2^3(5-1)+1][2^3(5-1)+2] = 561$ – the triangular number of rank 33 that was listed above when presenting the first five terms of the triangular number sieve of order three.

Input interpretation

$\text{Table}\left[2^{2k-1} n^2 - (2^{2k} - 3 \times 2^{k-1})n + (2^k - 1)(2^{k-1} - 1),\right.$
$\{n,$ arithmetic progression 1 to 10 step size $1\},$
$\left.\{k,$ arithmetic progression 0 to 9 step size $1\}\right]$

Result

1	1	1	1	1	1	1	1	1	1
3	6	15	45	153	561	2145	8385	33 153	131 841
6	15	45	153	561	2145	8385	33 153	131 841	525 825
10	28	91	325	1225	4753	18 721	74 305	296 065	1 181 953
15	45	153	561	2145	8385	33 153	131 841	525 825	2 100 225
21	66	231	861	3321	13 041	51 681	205 761	821 121	3 280 641
28	91	325	1225	4753	18 721	74 305	296 065	1 181 953	4 723 201
36	120	435	1653	6441	25 425	101 025	402 753	1 608 321	6 427 905
45	153	561	2145	8385	33 153	131 841	525 825	2 100 225	8 394 753
55	190	703	2701	10 585	41 905	166 753	665 281	2 657 665	10 623 745

Figure 7.10. Triangular number sieves of orders 0, 1, 2, ..., 10 vertically displayed.

$$\frac{(2^k\cdot(n-1)+1)\cdot(2^k\cdot(n-1)+2)}{2} - (2^{2\cdot k-1}\cdot n^2 - (2^{2\cdot k} - 3\cdot 2^{k-1})\cdot n + (2^k-1)\cdot(2^{k-1}-1))$$

$$\frac{(2^k\,(n-1)+1)\,(2^k\,(n-1)+2)}{2} - 2^{2k-1}n^2 + (2^{2k} - 3\,2^{-1+k})\,n - (2^k-1)\,(2^{-1+k}-1)$$

simplify(%)

$$0$$

Figure 7.11. Verifying numeric equivalence of (7.12) and (7.19) using Maple.

This exploration, focusing on the ranks of the elements of triangular number sieves of different orders, explains where the powers of two came from. Furthermore, using Maple (Figure 7.11) provides computational triangulation for the first problem-solving strategy (section 7.4) based on the use of Wolfram Alpha yielding formula (7.12).

7.6. The Third Problem-Solving Strategy of Constructing Triangular Number Sieves

One can also use a spreadsheet to generate triangular number sieves by using formulas describing the ranks of the terms of the sieves developed in the previous section. Formula (7.18) defines the n-th rank of the term of the sieve of order k. This general formula can be applied to other polygonal number sieves using formula (7.1) that defines an m-gonal number of rank n. For example, formula (7.19) defines the n-th term of the triangular number sieve of order k. The spreadsheet that generates triangular number sieves of orders zero through eight is shown in Figure 7.12. The third problem-solving strategy of constructing triangular number sieves makes it possible to generalize the spreadsheet-based triangular number sieve of order k to an m-gonal number sieve of order k using formula (7.1) representing an m-gonal number of order (rank) n. In that way, the formula

$$p_{n,m,k} = \frac{2^k(n-1)[2^k(n-1)+1](m-2)}{2} + 2^k(n-1) + 1 \tag{7.20}$$

entered in cell D4 and replicated to cell L16 with the value of m displayed in cell B2 (Figure 7.13) makes it possible to have a single spreadsheet that displays m-gonal number sieves of different orders. In particular, the spreadsheet of Figure 7.13 shows hexagonal ($m = 6$) number sieves of orders zero through eight. One can check to see that every other triangular number is a hexagonal number (Figure 7.13, column D beginning from cell D4).

Formula (7.20) represents the polygonal number sieve of order k and side m which rank is equal to $2^k(n-1)+1$. Using the spreadsheet of Figure 7.13, one can generate polygonal number sieves of different orders by changing the value of m in cell B2. In that way, formula (7.20) may be considered as a family of two-dimensional formulas depending on m.

	A	B	C	D	E	F	G	H	I	J	K	L	M	N
1		Order of sieve	0	1	2	3	4	5	6	7	8	9	10	11
2	Counting numbers	Powers of two	1	2	4	8	16	32	64	128	256	512	1024	2048
3	1	sieves	1	1	1	1	1	1	1	1	1	1	1	1
4	2		3	6	15	45	153	561	2145	8385	33153	131841	525825	2100225
5	3		6	15	45	153	561	2145	8385	33153	131841	525825	2100225	8394753
6	4		10	28	91	325	1225	4753	18721	74305	296065	1181953	4723201	18883585
7	5		15	45	153	561	2145	8385	33153	131841	525825	2100225	8394753	33566721
8	6		21	66	231	861	3321	13041	51681	205761	821121	3280641	13114881	52444161
9	7		28	91	325	1225	4753	18721	74305	296065	1181953	4723201	18883585	75515905
10	8		36	120	435	1653	6441	25425	101025	402753	1608321	6427905	25700865	102781953
11	9		45	153	561	2145	8385	33153	131841	525825	2100225	8394753	33566721	134242305
12	10		55	190	703	2701	10585	41905	166753	665281	2657665	10623745	42481153	169896961
13	11		66	231	861	3321	13041	51681	205761	821121	3280641	13114881	52444161	209745921
14	12		78	276	1035	4005	15753	62481	248865	993345	3969153	15868161	63455745	253789185
15	13		91	325	1225	4753	18721	74305	296065	1181953	4723201	18883585	75515905	302026753

Figure 7.12. Spreadsheet modeling of triangular number sieves of orders 0 through 11.

	A	B	C	D	E	F	G	H	I	J	K	L
1												
2	m	6	Order of sieve 0		1	2	3	4	5	6	7	8
3		Counting numbers	Powers of two 1		2	4	8	16	32	64	128	256
4		1	sieves	1	1	1	1	1	1	1	1	1
5		2		6	15	45	153	561	2145	8385	33153	131841
6		3		15	45	153	561	2145	8385	33153	131841	525825
7		4		28	91	325	1225	4753	18721	74305	296065	1181953
8		5		45	153	561	2145	8385	33153	131841	525825	2100225
9		6		66	231	861	3321	13041	51681	205761	821121	3280641
10		7		91	325	1225	4753	18721	74305	296065	1181953	4723201
11		8		120	435	1653	6441	25425	101025	402753	1608321	6427905
12		9		153	561	2145	8385	33153	131841	525825	2100225	8394753
13		10		190	703	2701	10585	41905	166753	665281	2657665	10623745
14		11		231	861	3321	13041	51681	205761	821121	3280641	13114881
15		12		276	1035	4005	15753	62481	248865	993345	3969153	15868161
16		13		325	1225	4753	18721	74305	296065	1181953	4723201	18883585

Figure 7.13. Spreadsheet modeling of hexagonal number sieves of orders 0 through 8.

Input interpretation

$$\text{Table}\left[2^k(n-1)(2^k(n-1)+1)\cdot\frac{m-2}{2}+2^k(n-1)+1,\right.$$

$\{k,$ arithmetic progression 0 to 4 step size 1 $\},$

$\{n,$ arithmetic progression 1 to 5 step size 1 $\},$

$\{m,$ arithmetic progression 3 to 7 step size 1 $\}]$

Result

Enlarge Data Customize Plain Text

{1, 1, 1, 1, 1}	{3, 4, 5, 6, 7}	{6, 9, 12, 15, 18}	{10, 16, 22, 28, 34}	{15, 25, 35, 45, 55}
{1, 1, 1, 1, 1}	{6, 9, 12, 15, 18}	{15, 25, 35, 45, 55}	{28, 49, 70, 91, 112}	{45, 81, 117, 153, 189}
{1, 1, 1, 1, 1}	{15, 25, 35, 45, 55}	{45, 81, 117, 153, 189}	{91, 169, 247, 325, 403}	{153, 289, 425, 561, 697}
{1, 1, 1, 1, 1}	{45, 81, 117, 153, 189}	{153, 289, 425, 561, 697}	{325, 625, 925, 1225, 1525}	{561, 1089, 1617, 2145, 2673}
{1, 1, 1, 1, 1}	{153, 289, 425, 561, 697}	{561, 1089, 1617, 2145, 2673}	{1225, 2401, 3577, 4753, 5929}	{2145, 4225, 6305, 8385, 10465}

Figure 7.14. Wolfram Alpha's use of formula (7.20) as a three-dimensional model.

At the same time, the use of Wolfram Alpha makes it possible to use formula (7.20) as a three-dimensional modeling tool by creating a table of the elements of the corresponding polygonal number sieves for different values of m. Such table is shown at the bottom part of Figure 7.14 generating the first five elements of the five sieves for $n \in [1,5], k \in [0,4], m \in [3,7]$. In this table, horizontally, the first position in each row contains numbers that belong to the triangular number sieves of orders zero through four. For example, the first element in the second bracket of the fifth row of the table (Figure 7.14) is 153. This number can be found in cell G4 of the spreadsheet of Figure 7.12 – the second element of the triangular number sieve of order four. Likewise, the second, third, fourth and fifth positions in each row contain numbers that belong, respectively, to square, pentagonal, hexagonal, and heptagonal number sieves of orders zero through four. For example, the fourth element in the third bracket of the fifth row of the table (Figure 7.14) is 2145. This number can be found in cell H6 of the spreadsheet of Figure 7.13 – the third element of the hexagonal number sieve of order four. In that way, the joint use of Wolfram Alpha and a spreadsheet provides computational triangulation in the context of constructing polygonal number sieves of different sides and orders. Furthermore, as will be discussed in the next section, the table of Figure 7.14 generated by Wolfram Alpha prompts the development of other types of sieves of polygonal numbers.

7.7. The Fourth Problem-Solving Strategy Leading to Cullen-Type Polygonal Number Sieves

Explorations of this section begin with interpreting the results of computations described in the previous three sections in mathematical terms, using the TITE framework (Chapter 3, Section 3.5). However, while a TI part of explorations will follow the TE part provided by a spreadsheet and Wolfram Alpha, the former part will not be totally independent of technology by seeking information available online and carrying out the very basic computation of products and exponents as a verification of formal reasoning through falling back on the results of digital computations. Consider the numbers 1, 6, 45, 325, 2145, which, according to the table of Figure 7.14 represent the 1st, 2nd, 3rd, 4th, and 5th elements of triangular number sieves of orders 0, 1, 2, 3, and 4, respectively. In particular, the number 2145 can be found in cell G7 of the spreadsheet of Figure 7.12 as the 5th element of the triangular number sieve of order 4 (alternatively, the 5th sieve, if the sieve of order zero is considered the 1st sieve). The ranks of the above five triangular numbers are 1, 3, 9, 25, 65, the general form of which, $r(n) = (n-1) \cdot 2^{n-1} + 1, n = 1, 2, 3, \ldots$, can be found in the OEIS® (A002064) where they are referred to as Cullen numbers, named after James Cullen (1867-1933), an Irish mathematician and Jesuit priest. Indeed, $r(1) = 1$, $r(2) = 3$, $r(3) = 9$, $r(4) = 25$, $r(5) = 65$, and so on.

One can check to see that $r(6) = 161$, so that the 6th element of the 6th triangular number sieve (assigning $n = 1$ to the sieve of order zero) is $\frac{161 \cdot 162}{2} = 13041$ (see cell H8 in the spreadsheet of Figure 7.12). Also, $r(7) = 385$ and the 7th element of the 7th triangular number sieve is $\frac{385 \cdot 386}{2} = 74305$ (see cell I9 in the spreadsheet of Figure 7.12). In order to define the sequence of triangular numbers 1, 6, 45, 325, 2145, 13041, 74305, ..., (which can be interpreted as the Cullen-type triangular number sieve because its elements have the ranks $r(1) = 1$, $r(2) = 3$, $r(3) = 9$, $r(4) = 25$, $r(5) = 65$, $r(6) = 161$, $r(7) = 385$), one has to use the formula

$$t_{r(n)} = \frac{[(n-1)\cdot 2^{n-1}+1][(n-1)\cdot 2^{n-1}+2]}{2}, n = 1, 2, 3 \ldots . \quad (7.21)$$

For example, $t_{r(6)} = \frac{(5\cdot 2^5+1)(5\cdot 2^5+2)}{2} = \frac{161\cdot 162}{2} = 13041$.

Alternatively, the sequence $t_{r(n)}$ defined by formula (7.21) can be described through the following action. To move along the triangular number sieve from 1 to 6, one skips 1 triangular number (3); to move from 6 to 45 one has to skip 5 triangular numbers (10, 15, 21, 28, 36); to move from 45 to 325 one has to skip 15 triangular numbers; to move from 325 to 2145 one has to skip 39 triangular numbers; to move from 2145 to 13041 one has to skip 95 triangular numbers, to move from 13041 to 74305 one has to skip 225 triangular numbers. The general form of the sequence 1, 5, 15, 39, 95, 225 is $a(n) = (n+1)\cdot 2^{n-1} - 1, n = 1,2,3,...$ (e.g., $a(1) = 1$, $a(2) = 5$) and it can be found in the OEIS® under the number A099035. That is, the Cullen-type triangular number sieve 1, 6, 45, 325, 2145, 13041 can be defined as the step-by-step elimination of $(n+1)\cdot 2^{n-1} - 1$ triangular numbers. This elimination yields triangular numbers of the ranks $n \cdot 2^n + 1$, $n \geq 1$. Indeed, when $n = 1$ (the first step of elimination) we have $[(n+1)\cdot 2^{n-1} - 1]|_{n=1} = 1$ – meaning the elimination of one triangular number, 3, and retaining the triangular number 6 the rank of which is $(n \cdot 2^n + 1)|_{n=1} = 3$. When $n = 2$ (the second step of elimination) we have $[(n+1)\cdot 2^{n-1} - 1]|_{n=2} = 5$ – meaning the elimination of five triangular numbers, (10, 15, 21, 28, 36), retaining the triangular number 45 the rank of which is $(n \cdot 2^n + 1)|_{n=2} = 9$.

Likewise, one can consider the numbers 1, 15, 153, 1225, 8385, which, according to the table of Figure 7.14, represent the 1st, 2nd, 3rd, 4th, and 5th elements of hexagonal number sieves of orders 0, 1, 2, 3, and 4, respectively. In particular, the number 8385 can be found in cell H8 of the spreadsheet of Figure 7.13 as the 5th element of the 5th hexagonal number sieve. The ranks of the above five numbers are also Cullen numbers $r(n) = (n-1)\cdot 2^{n-1} + 1, n = 1,2,3,...$. Thus, as $r(6) = 161$, the 6th element of the 6th hexagonal number sieve is $161 \cdot (2 \cdot 161 - 1) = 51681$ (see cell I9 in the spreadsheet of Figure 7.13 and formula (7.4)). In order to define the sequence of hexagonal numbers 1, 15, 153, 1225, 8385, 51681, ... , which can be interpreted as the Cullen-type hexagonal number sieve, one has to use the formula

$$h_{r(n)} = [(n-1)\cdot 2^{n-1} + 1]\cdot\left[2\cdot\left((n-1)\cdot 2^{n-1} + 1\right) - 1\right], n = 1,2,3 \ldots . \qquad (7.22)$$

For example, $h_{r(6)} = (5 \cdot 2^5 + 1)\cdot(2\cdot(5\cdot 2^5 + 1) - 1) = 161 \cdot 321 = 51681$ (see cell I9 in the spreadsheet of Figure 7.13 and formula (7.4)).

Generalizing from formulas (7.21) and (7.22), the following formula

$$P_{r(n),m} = (n-1)\cdot 2^{n-2}\cdot[(n-1)\cdot 2^{n-1} + 1]\cdot(m-2) + (n-1)\cdot 2^{n-1} + 1, \qquad (7.23)$$

where $n \geq 1, m \geq 3$, for the Cullen-type polygonal number sieve can be derived. For example, setting $n = 3$ and $m = 4$ in formula (7.23) yields

$P_{r(3),4} = (3-1) \cdot 2^{3-2} \cdot [(3-1) \cdot 2^{3-1} + 1] \cdot (4-2) + (3-1) \cdot 2^{3-1} + 1 = 81$ – the square number ($m = 4$) of rank $r(3) = 9$.

Input interpretation

Table[$(n-1) \times 2^{n-2} ((n-1) \times 2^{n-1} + 1)(m-2) + (n-1) \times 2^{n-1} + 1$,
{n, arithmetic progression 1 to 9 step size 1},
{m, arithmetic progression 3 to 11 step size 1}]

Result

1	1	1	1	1	1	1	1	1
6	9	12	15	18	21	24	27	30
45	81	117	153	189	225	261	297	333
325	625	925	1225	1525	1825	2125	2425	2725
2145	4225	6305	8385	10465	12545	14625	16705	18785
13041	25921	38801	51681	64561	77441	90321	103201	116081
74305	148225	222145	296065	369985	443905	517825	591745	665665
402753	804609	1206465	1608321	2010177	2412033	2813889	3215745	3617601
2100225	4198401	6296577	8394753	10492929	12591105	14689281	16787457	18885633

Figure 7.15. Cullen-type m-gonal number sieves after the first elimination ($3 \leq m \leq 11$).

The table (a 9×9 matrix) of Figure 7.15 created by Wolfram Alpha allows one to write down the first three terms of the Cullen-type number sieves after the second elimination for triangular numbers (1, 45, 2100225), square numbers (1, 81, 4198401), pentagonal numbers (1, 117, 6296577), hexagonal numbers (1, 153, 8394753), and so on. One can check to see that the third element in each of the four triples has rank 2049 on the original list of the corresponding polygonal numbers. For example, the original ranks of triangular, 2100225, and square, 4198401, numbers prior to any elimination are $\frac{-1+\sqrt{1+8 \cdot 2100225}}{2} = 2049$, as it follows from formula (7.2), and $\sqrt{4198401} = 2049$, respectively. Furthermore, those third elements, respectively, have rank 9 (as $[(n-1) \cdot 2^{n-1} + 1]|_{n=3} = 9$) on the list of the corresponding triangular, square, pentagonal, and hexagonal number sieves of order eight (alternatively, the 9th sieve, counting the sieve of order zero as the 1st sieve). Likewise, the triangular number 45 and the hexagonal number 153 appear in cells F4 and G5 of the spreadsheets of Figure 7.12 and Figure 7.13, respectively, both having rank 3 (as $[(n-1)\cdot 2^{n-1}+1]|_{n=2}=3$) on the lists of the corresponding 4th sieves (including sieves of order zero). One can check to see that the same is true for the square, 81, and the pentagonal, 117, numbers – both have rank 3 on the list of corresponding 4th sieves (counting the sieve of order zero as the first sieve). That is, regardless of the side of polygonal numbers, the Cullen-type elimination process provides the same place within a sieve for the same-rank survivors.

To explain mathematical meaning of such connections, note that among the first n natural numbers there are $INT(\frac{n}{2})$ even numbers and, because (setting $n = 2l$ and $n = 2l + 1$, depending on the parity of n)

$$n - INT\left(\frac{n}{2}\right) = \begin{cases} 2l - INT\left(\frac{2l}{2}\right) = l \\ 2l + 1 - INT\left(\frac{2l+1}{2}\right) = l + 1 \end{cases},$$

there are $INT\left(\frac{n}{2}\right) + 1$ numbers that survive elimination of every other number from a polygonal number sieve (whatever its number). In other words, if a polygonal number in a sieve has rank n, on the next elimination step, when $n = 2l$ it has rank l, and it has rank $l + 1$ when $n = 2l + 1$. That is, if a number survives elimination, its rank in the corresponding sieve decreases. Therefore, the rank 2049 of a polygonal number on the original list (e.g., $2100225 = \frac{2049\cdot(2049+1)}{2}$ on the list of triangular numbers) would follow the following transformations before this number appears in the 3rd sieve:

$$2049 \to (2049+1) \div 2 = 1025 \to (1025+1) \div 2 = 513 \to (513+1) \div 2 = 257 \to (257+1) \div 2 = 129 \to (129+1) \div 2 = 65 \to (65+1) \div 2 = 33 \to (33+1) \div 2 = 17 \to (17+1) \div 2 = 9 \to (9+1) \div 2 = 5.$$

That is, each number in this chain of ranks beginning from the second is equal to the previous number increased by one and divided by two. For example, the triangular number 2100225 ($= 2049 \cdot (2049+1)/2$) appearing in the 12th sieve (counting the zero sieve, Figure 7.12, cell N4) has rank 2049 and the triangular number 15 ($= 5 \cdot (5+1)/2$) appearing in the 3rd sieve (counting the zero sieve, Figure 7.12, cell E4) has rank 5.

One can also start from rank 2049 and move up to determine ranks of the polygonal numbers and their corresponding values by decreasing by one the double of each rank to get the following chain of ranks:

$2049 \leftarrow (2 \cdot 2049 - 1) = 4097 \leftarrow (2 \cdot 4097 - 1) = 8193 \leftarrow (2 \cdot 8193 - 1) = 16385 \leftarrow (2 \cdot 16385 - 1) = 32769 \leftarrow (2 \cdot 32769 - 1) = 65537 \leftarrow (2 \cdot 65537 - 1) = 131073 \leftarrow (2 \cdot 131073 - 1) = 262145 \leftarrow (2 \cdot 262145 - 1) = 524289.$

That is, each number in this chain of ranks beginning from the second is twice the previous number decreased by one. Using formulas connecting polygonal numbers to their ranks, the corresponding polygonal numbers can be found. For example, $t_{4097} = \frac{4097 \cdot (4097+1)}{2} = 8394753$ and $t_{8193} = \frac{8193 \cdot (8193+1)}{2} = 33566721$. These two triangular numbers can be seen in cells L11 and M11 of the spreadsheet of Figure 7.12 – they both have rank 9 (cell A11), but belong to the 9th and the 10th sieves, respectively.

One can see how computations provide learners of mathematics with numeric evidence for recognizing patterns which require explanations that are independent from the use of technology (i.e., technology-immune). At the same time, when mathematical results are driven by computations (i.e., technology-enabled), formal reasoning may not be entirely detached from computing as reasoning constantly resorts to the results of digital computations for the verification of the absence of possible errors. In other words, computation is the primary instrument used to trigger, amplify, and verify mathematical argument.

Conclusion

The chapter used computational triangulation to explore several subsequences of natural numbers using the TITE framework. It started with the exploration of triangular number sieves of different orders using three problem-solving strategies. The explorations were extended to introduce polygonal number sieves of different orders and to develop the general formula that define such sieves. The chapter concluded with the introduction of polygonal number sieves of the Cullen type. The material of the chapter was enhanced by different references to the history of mathematics associated with the representation of integers as sums of other integers. The scope of this chapter is appropriate to be included in a course for prospective secondary mathematics teachers. The next chapter will use the TITE framework and computational triangulation when exploring two discrete mathematics problems with historical flavor.

Chapter 8

TITE Explorations of Two Discrete Mathematics Topics with Historical Flavor

8.1. Introduction

In this final chapter, two topics from discrete mathematics appropriate for advanced exploration with future teachers of secondary mathematics will be discussed using the TITE framework (Chapter 3, Section 3.5). The first topic deals with the use of computational triangulation in computing the probability of the irreducibility of a fraction with randomly selected numerator and denominator from the set of natural numbers. This topic, discussed from the computational perspective elsewhere (Abramovich & Nikitin, 2017), concerns historically famous problem situated at the confluence of the theories of numbers and probability. The problem is associated with several notable figures in the history of mathematics including a German mathematician Peter Gustav Lejeune Dirichlet (1805-1859), a Russian mathematician Pafnuty Lvovich Chebyshev (1821-1894), an Italian mathematician Ernesto Cesàro (1859-1906) and was even referred to as a theorem of Gauss (Arnold, 2003, 2015). In this regard note that a German mathematician Felix Klein (1849-1925), the first president of the International Commission on Mathematical Instruction, in his *Elementary Mathematics from a Higher Standpoint*, ascribed great importance to the use of history of mathematics in the teaching of the subject matter (Klein, 2016).

The second topic shows how turning the symmetry of Pascal's triangle into the asymmetry of its representation with staggered rows makes it possible, by combining computational triangulation and the TITE framework, to develop a special class of one-variable polynomials, called Fibonacci-like polynomials, the coefficients of which add up to Fibonacci numbers, two polynomials in each degree, and all roots of which are real numbers. Within each degree, the roots alternate their location on the number line between symmetry and asymmetry. Fibonacci-like polynomials stem both from Pascal's triangle and from defining the cyclic behavior of iterations of a two-parametric non-linear difference equation in terms of algebraic relations involving continued fractions and leading to the notion of generalized Golden Ratios in the form of cycles as strings of numbers of different lengths.

8.2. From Spinner to Euler's Identity

Experimental character of computing the probability of the irreducibility of a fraction the numerator and denominator of which are chosen at random from the set of natural numbers requires replacing probability by relative frequency to avoid the use of mathematical machinery studied at the tertiary level by future professional mathematicians. Three digital tools will be used in computations – a spreadsheet, Wolfram Alpha, and Maple – enabling computational triangulation of the results. In addition, the Graphing Calculator will be used for graphing and GeoGebra for explaining computations to be carried out in the contexts of Wolfram Alpha and Maple.

To demonstrate the elementary origin of the problem, one can begin with the following simple technology-immune (TI) inquiry: What is the probability of landing on an even number of a spinner divided into eight equal sections and labeled by the first eight natural numbers? The answer 1/2 can be found as "the ratio of the number of favorable cases to that of all possible cases … that supposes the various cases equally likely" (Laplace, 1814/1951, p. 11). One assumes that the outcomes of landing on one of the eight equal sections of the spinner are equally likely, making it possible to use the above definition of probability by Pierre Simone Laplace (1749-1827, France) who is regarded as one of the greatest scholars of all time, being a notable contributor to probability theory. Under the same assumption, when spinner is divided into nine equal sections, the chances for an even number (i.e., four out of nine) become smaller than 1/2 as 4/9 < 4/8 = 1/2. Yet, with ten sections, the chances (five out of ten) are back to 1/2. Now, the following question can be asked: What is the probability of randomly selecting an even number from a set of consecutive natural numbers?

To answer this question in a TI way, one must know the cardinality of such a set. With the cardinality measured by an even number, the probability is 1/2. For example, the range [10, 29] includes 20 integers, half of which are even numbers; in general, among 2n consecutive natural numbers, there are n even numbers. When the cardinality of a set is an odd number, the probability (understood as relative frequency) of selecting an even number from this set is equal to $\frac{n}{2n+1}$ where the denominator is the cardinality of the set, and the numerator is the number of even numbers within the set. Due to the inequalities $\frac{n}{2n+1} < \frac{n+1}{2n+3} < \frac{1}{2}$ (that hold true for all n ≥ 1, as the cross multiplication yields $2n^2 + 3n < 2n^2 + 3n + 1$ whence $0 < 1$), one can conclude that the relative frequency of selecting an even number from the first 2n + 1 natural numbers, by increasing monotonically, becomes closer and closer to 1/2 (see Figure 8.1). One can also select the multiples of, say, 5 and, using the fraction 1/5 to measure the chances of such selection through a technology-enabled (TE) technique. The spreadsheet of Figure 8.2 shows that relative frequency of selecting multiples of 5 among 10,000 randomly generated integers in the range $[2, 10^4]$ is equal to $0.1967 \cong 0.2 = 1/5$ (cell D1). The result (which slightly changes with a new set of 10,000 randomly generated integers) is then compared to 1/5 showing the difference as 0.0033 (cell F1). Likewise, one can heuristically, without rigorous justification, say that the unit fraction $1/p$ represents the probability of selecting the prime number p from the set of natural numbers. Furthermore, the probability that the common fraction $\frac{a}{b}$ is reducible by 5 is equal to $\frac{1}{25}$. Consequently, the probability that the fraction $\frac{a}{b}$ is not reducible by 5 is equal to $1 - \frac{1}{25} = \frac{24}{25}$. Likewise, the probability that the fraction $\frac{a}{b}$ is not reducible by 2 is equal to $1 - \frac{1}{4} = \frac{3}{4}$. Therefore, the product $\frac{24}{25} \cdot \frac{3}{4} = \frac{18}{25} = 0.72$ is the probability that the fraction $\frac{a}{b}$ is not reducible by either 2 or 5. Note that this probability does not change significantly when irreducibility by, say, 23 (another prime number) is considered as well, because $1 - \frac{1}{23^2} \cong 0.998$ and the product $0.72 \cdot 0.998 \cong 0.719$ differs from the first factor by 0.001.

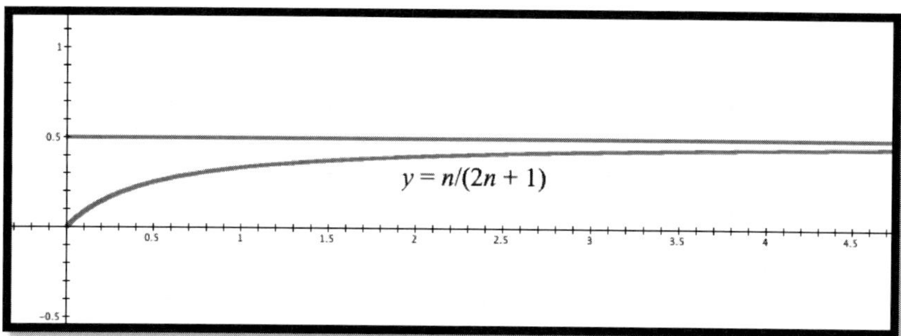

Figure 8.1. The fraction $n/(2n + 1)$ approaches 1/2 monotonically increasing.

	A	B	C	D	E	F
1		5		0.1967		0.0033
2	3335	1				
3	8995	1				
4	3338					
5	4057					
10000	827					
10001	8139					
10002						

Figure 8.2. A relative frequency of the number 5 among 10,000 integers is close to 1/5.

To demonstrate purely mathematical approach to finding the probability of irreducibility of a fraction with randomly selected natural numbers as its numerator and denominator, that is, to use a TI component of the activities, one has to heuristically (i.e., without rigorous justification) make the following three statements. First, the probability of selecting two multiples of the prime number p from the set of all natural numbers is $\frac{1}{p^2}$. Indeed, the two selections, like in the case of tossing two coins or rolling two dice, do not depend on each other. Second, the probability of randomly selecting two natural numbers not having the prime number p as a common factor is equal to $1-\frac{1}{p^2}$. Third, due to the second statement (and the above numerical evidence in the case of the product $\left(1-\frac{1}{2^2}\right)\left(1-\frac{1}{5^2}\right)\left(1-\frac{1}{23^2}\right)$ of three probabilities), the probability of irreducibility of a common fraction is equal to the infinite product $\prod_p(1-\frac{1}{p^2})$ over all prime numbers p. This product is connected to the famous Euler's identity

$$\prod_p(1-\frac{1}{p^2})^{-1}=\sum_{n=1}^{\infty}\frac{1}{n^2}=\frac{\pi^2}{6}$$

from where it follows that

$$\prod_p\left(1-\frac{1}{p^2}\right)=\frac{6}{\pi^2}\cong 0.608.$$

In the next section, the number 0.608 will be confirmed by using different informal approaches supported by digital tools as a TE part of the activities. In particular, computational triangulation will make it possible to demonstrate "the empirical character of the theory of probability and its purpose of interpreting observable phenomena" (von Mises, 1957, p. 64).

8.3. Computational Triangulation on the Problem of Irreducibility of Fractions

One can further enable technology to approach the problem of irreducibility of fractions experimentally and, by using a spreadsheet, randomly generate 10,000 pairs of natural numbers in the range $[1, 10^4]$. Figure 8.3 shows the spreadsheet displaying in columns A and B the first five and the last two pairs only; in column C the tool computes for each pair their greatest common divisor (GCD) and the number of cells with the unity as the sought value of the GCD is computed in cell D2. Next, the spreadsheet calculates experimental probability (relative frequency) of the co-primality of two natural numbers out of randomly generated 10,000 pairs (in other words, carrying out 10^4 trials of co-primality), then each such trial is repeated 500 times (by changing the value of cell G2 through a slider attached to that cell), thereby making the total number of trials $5 \cdot 10^6$. The value 0.608126 as the average over 500 trials is displayed in cell G4. In the spirit of computational triangulation, this number will be confirmed by Wolfram Alpha and Maple. Finally, the results of computational triangulation will be compared with $\frac{6}{\pi^2}$ – the theoretical probability of irreducibility of a common fraction with randomly selected numerator and denominator from the set of natural numbers.

	A	B	C	D	E	F	G
1	a	b	GCD(a, b)	GCD(a, b)=1		Exp. Probability	
2	1697	4221	1	6054	1	0.603360336	500
3	8717	310	1		2	0.602960296	
4	414	7030	2		3	0.614461446	0.608126
5	626	7276	2		4	0.609460946	
6	6137	4915	1		5	0.608260826	
9999	3913	3595	1				
10000	7735	3911	1				

Figure 8.3. Confirming the number 0.608 through five million trials.

To use other digital tools, Wolfram Alpha and Maple, one can start with interpreting the problem of irreducibility of common fractions geometrically. To this end, consider a lattice plane (known to young children and their teachers as a geoboard) – a rectangular grid in the form of the periodic array of (lattice) points located in the first quadrant of the coordinate plane. On such a grid, one can locate points the coordinates of which are relatively prime numbers. The process of location is purely geometrical – a segment connecting the origin to a point with relatively prime coordinates must be free from other lattice points. Indeed, if another lattice point belongs to this segment, due to the similarity of triangles (discussed in Chapter 1), the end point of the segment would not have relatively prime coordinates because of the proportion $\frac{x_1}{y_1} = \frac{x_2}{y_2}$, where (x_1, y_1) and (x_2, y_2) are, respectively, the coordinates of the end point and another lattice point through which the segment passes and, therefore, this proportion shows reducibility of the fraction $\frac{x_1}{y_1}$. Figure 8.4 created by the Graphing Calculator shows two points, (3, 5) and (6, 4), connected to the origin. The former point belongs to the segment free from any lattice point (except the origin); the latter point belongs to the segment passing through the point (3, 2) as $\frac{6}{4} = \frac{3}{2}$. One can define a lattice point as representing an irreducible fraction if the segment connecting this point with the origin is free from other lattice points. According to this definition, among the lattice points of the first quadrant, the points (1, 0) and (0, 1) represent irreducible fractions as well.

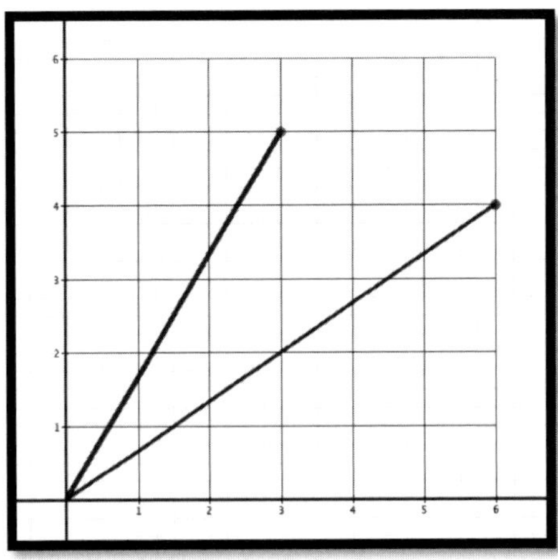

Figure 8.4. Geometric interpretation of irreducibility of fractions.

Figure 8.5 created in GeoGebra shows all lattice points within the 6×6 square having the origin as its bottom-left corner. One can count the number of such points one by one. There are 5 points within the 2×2 square; 4, 4, 8, and 4 points on the top and far-right borders of the $3 \times 3, 4 \times 4, 5 \times 5$ and 6×6 squares, respectively. That is, the total number of points representing irreducible fractions is $5 + 4 + 4 + 8 + 4 = 25$, among the total of $35\ (= 6^2 - 1)$ lattice points. The ratio $\frac{25}{35} \cong 0.71$ (compare this ratio to the number 0.719 mentioned in Section 8.2). One can also see that among those 25 points, the number of points on the far-right (or top) border of the 3×3 square is 2 and the number of positive integers smaller than and relatively prime to 3 is 2 (i.e., 1 and 2). Likewise, the number of points on the far-right (or top) border of the 4×4 square is 2 and the number of positive integers smaller than and relatively prime to 4 is 2 (i.e., 1 and 3). The same description can be applied to the numbers 4, and 2 associated with the points on the far-right (or top) borders of the 5×5 and 6×6 squares.

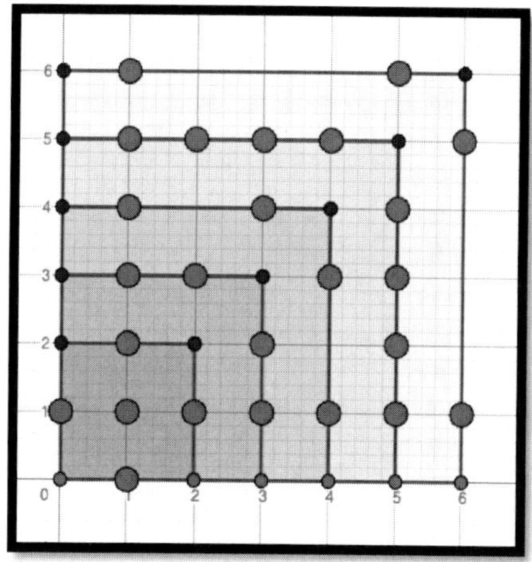

Figure 8.5. Large lattice points representing irreducible fractions within the 6×6 square.

In general, moving from counting to computing (Abramovich, 2025a), whatever the number of lattice points representing irreducible fractions can be found within the $(n-1) \times (n-1)$ square, the transition to the $n \times n$ square augments that number by twice the number of positive integers smaller than and relatively prime to n.

Input interpretation

$$\sum_{2}^{1000} 2\,\phi(n)$$

$$\sum_{n=2}^{1000} 2\,\phi(n) = 608\,382$$

Figure 8.6. Using Wolfram Alpha in computing Euler's phi function.

Input

$$\frac{608\,382 + 3}{1000^2 - 1}$$

Exact result

$$\frac{202\,795}{333\,333}$$

Decimal approximation

0.60838560838560838560838560838560838560838560838560838560838560…
8385608386

Figure 8.7. Confirming the number 0.608 using Wolfram Alpha.

with(NumberTheory);

$sum(2 \cdot \text{phi}(n), n = 2..1000);$

608382

$$\frac{(3 + \%)}{1000^2 - 1}$$

0.60838860838860838861

Figure 8.8. Confirming the number 0.608 using Maple.

Such augmentation can be represented by Euler's phi (alternatively, totient) function $\varphi(n)$ which returns the number of positive integers smaller than and relatively prime to n. In order to find the number of lattice points within the $n \times n$ square (having the origin as its bottom-left corner), one can compute the value of $3 + \sum_{i=2}^{n} 2\varphi(i)$ by using Wolfram Alpha for $n = 1{,}000$ (Figure 8.6). The total number of all lattice points within the $n \times n$ square (excluding the origin) is equal to $n^2 - 1$ and, therefore, the relative frequency of irreducibility of a common fraction with the numerator and the denominator selected from the range $[1, n]$ is equal to $\frac{3+\sum_{i=2}^{n} 2\varphi(i)}{n^2-1}$. Using Wolfram Alpha (Figure 8.7) and Maple (Figure 8.8) the last fraction is computed for $n = 1{,}000$ yielding the result obtained theoretically using Euler's identity that led to the value $\frac{6}{\pi^2} \cong 0.608$. This concludes explorations of the problem with rich history that nowadays can be used for a demonstration of computational triangulation at the confluence of probability theory and theory of numbers.

8.4. The Asymmetry of Entries of Pascal's Triangle Yields Fibonacci-Like Polynomials

Several educational documents emphasize the importance of symmetry, and its duality often referred to as non-symmetry (alternatively, asymmetry). For example, in the United States, Common Core State Standards (2010) mention the word symmetry in the first grade (p. 13), in the second grade (p. 17) and in secondary school (p. 69) mathematics. In Canada, students in grade two learn that "shapes can be sorted by comparing geometric attributes such as ... the number of lines of symmetry ... [whereas] some triangles ... have no lines of symmetry" (Ontario Ministry of Education, 2020, pp. 168, 169) and in grade six are expected to "create lists of geometric properties ... including ... rotational symmetry, and line symmetry" (ibid, p. 367). In Singapore, at the primary school level, "in teaching the topic on Symmetry, teacher first shows two groups of shapes – symmetrical and non-symmetrical shapes ... guides students to focus on the attributes of symmetrical shapes" (Ministry of Education Singapore, 2012, p. 24). At the secondary level, Geometry and Measurement curriculum (Ministry of Education Singapore, 2020) includes teaching "properties of triangles ... including symmetry properties" (p. 16), symmetry of functions and graphs (p. 17), and "symmetry properties of circles" (p. 20).

Supported by the above-mentioned standards, the second topic discussed in this chapter through the lens of computational triangulation deals with the concepts of symmetry and asymmetry in the context of famous Pascal's triangle used by a French mathematician and philosopher Blaise Pascal (1623-1662) for recording sample spaces of experiments of tossing different number of coins (Kline, 1985). Other topics in school mathematics can be associated with symmetry and asymmetry (Abramovich, 2025b). Consider Pascal's triangle shown in Figure 8.9. One can see that the entries of Pascal's triangle, being binomial coefficients, are symmetrical within each row. The top right-bottom left diagonals of the triangle consist, respectively, of units, natural numbers (partial sums of units), triangular numbers (partial sums of natural numbers), triangular pyramidal numbers (partial sums of triangular numbers), pentatope numbers (partial sums of triangular pyramidal numbers), and so on (see also Chapter 6, Sections 6.1 and 6.5; Chapter 7, Section 7.2). Figure 8.10 shows asymmetrical rearrangement of those numbers when the diagonal with units forms the first column, the diagonal with natural numbers forms the second column shifted down about the first one by two rows, the diagonal with triangular numbers becomes the third column shifted down about the second one by two rows, the diagonal with triangular pyramidal numbers forms the fourth column shifted down about the third one by two rows, and so on.

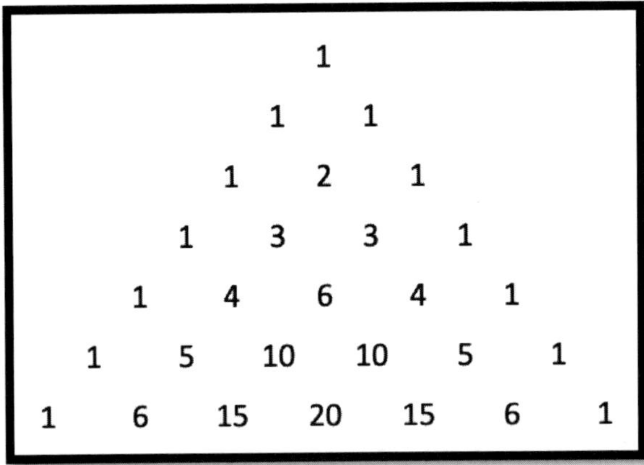

Figure 8.9. Pascal's triangle is symmetrical.

An interesting property of numbers appearing in each row of the modified Pascal's triangle (Figure 8.10), in addition to having their sums in each row equal to consecutive Fibonacci numbers (indeed, beginning from the third row: 1 + 1 = 2, 1 + 2 = 3, 1 + 3 + 1 = 5, 1 + 4 + 3 = 8, 1 + 5 + 6 + 1 = 13, and so on), is that using those numbers as coefficients of one-variable polynomials, results in polynomials all roots of which are real numbers located within the interval $(-4, 0)$. Furthermore, out of two polynomials of the same degree, one polynomial has symmetrical location of roots, and another polynomial has asymmetrical location of roots. These observations would not be possible without the use of digital technology. For example, the numbers 1, 6, 10, 4, the sum of which is the 8th Fibonacci number 21, form the polynomial $f(x) = x^3 + 6x^2 + 10x + 4$ the graph of which (Figure 8.11, left) intersects the interval $(-4, 0)$ three times. The two roots, $x = -2 - \sqrt{2}$ and $x = -2 + \sqrt{2}$ are symmetrical about the third root $x = -2$, which is the midpoint between the other two roots as $\frac{(-2-\sqrt{2})+(-2+\sqrt{2})}{2} = -2$. At the same time, the numbers 1, 5, 6, 1, the sum of which is the 7th Fibonacci number 13, form the polynomial $f(x) = x^3 + 5x^2 + 6x + 1$ the graph of which (Figure 8.11, right) intersects the interval $(-4, 0)$ three times as well, yet the points of intersection are not symmetrical as the following relations demonstrate: $-1.555 - (-3.247) = 1.642 \neq 1.357 = -0.198 - (-1.555)$.

Alternatively, the root -1.555 is not the midpoint between the other two roots as $\frac{(-3.247)+(-0.19806)}{2} = -1.72253$.

1					
1					
1	1				
1	2				
1	3	1			
1	4	3			
1	5	6	1		
1	6	10	4		
1	7	15	10	1	
1	8	21	20	5	
1	9	28	35	15	1
1	10	36	56	35	6

Figure 8.10. The modified Pascal's triangle is asymmetrical.

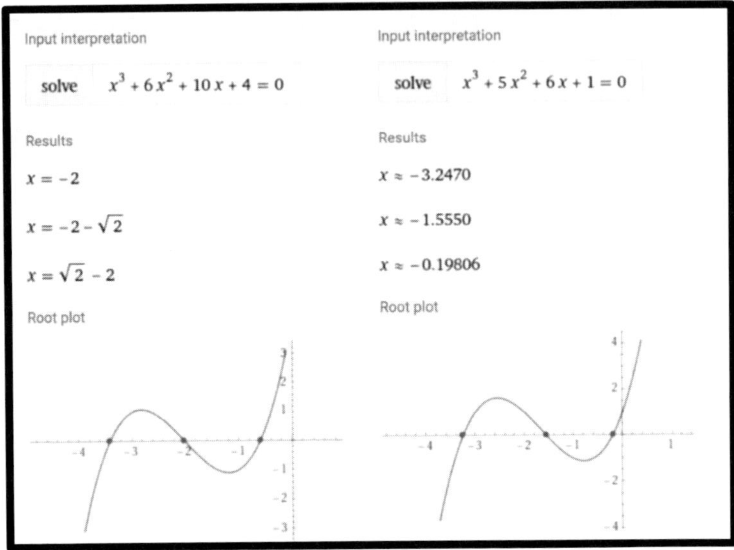

Figure 8.11. Fibonacci-like polynomials of degree three and their roots in Wolfram Alpha.

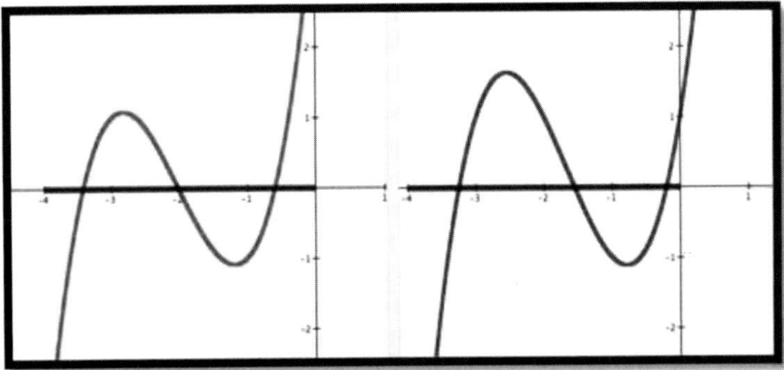

Figure 8.12. Using the Graphing Calculator to plot Fibonacci-like polynomials of degree three.

The notion of computational triangulation is applicable to computer graphing also. Using the Graphing Calculator, one can duplicate the results of Wolfram Alpha. Indeed, the graph of the Fibonacci-like polynomial $f(x) = x^3 + 6x^2 + 10x + 4$ (Figure 8.12, left) shows three intersections of the interval $(-4,0)$ displaying symmetry of the largest and the smallest roots about the third root. The graph of the polynomial $f(x) = x^3 + 5x^2 + 6x + 1$ (Figure 8.12, right) also shows three intersections displaying asymmetrical location of the roots. However, unlike Wolfram Alpha's automatic marking of intersections (roots), such option is not offered by the Graphing Calculator and marking of the roots requires additional digital fabrication of the roots as the intersection points. In Figure 8.12 the interval $(-4,0)$ is digitally fabricated through the system of inequalities $-4 < x < 0, |y| < \varepsilon$, for a relatively small value of ε (the 'thickness' of the interval). The issue of digital fabrication of points was discussed in Chapter 3, Section 3.4 and shown there in Figure 3.4.

Likewise, the numbers 1, 8, 21, 20, 5, the sum of which is the 10th Fibonacci number 55, form the polynomial $f(x) = x^4 + 8x^3 + 21x^2 + 20x + 5$ the graph of which (Figure 8.13, left) intersects the interval $(-4, 0)$ four times. The roots, $x = \frac{-5-\sqrt{5}}{2}$ and $x = \frac{\sqrt{5}-3}{2}$ are symmetrical about the point $x = -2$, as $-2 - \frac{-5-\sqrt{5}}{2} = \frac{1+\sqrt{5}}{2}$ and $\frac{\sqrt{5}-3}{2} - (-2) = \frac{1+\sqrt{5}}{2}$. The same is true for another pair of roots, as $\frac{\frac{-5-\sqrt{5}}{2}+\frac{\sqrt{5}-3}{2}}{2} = -2$. The numbers 1, 7, 15, 10, 1, the sum of which is the 9th Fibonacci number 34, form the polynomial $f(x) = x^4 + 7x^3 + 15x^2 + 10x + 1$ the graph of which (Figure 8.13, right) intersects the interval $(-4, 0)$ four times as well, yet the points of intersection clearly display asymmetrical location, because each pair of points is symmetrical about their own midpoint, not common for all the four roots.

In the spirit of computational triangulation, Figure 8.14 (Wolfram Alpha) and Figure 8.15 (Maple) display two polynomials of degree five, with coefficients borrowed from the last two rows of the modified Pascal's triangle of Figure 8.10, one of which, $f(x) = x^5 + 10x^4 + 36x^3 + 56x^2 + 35x + 6$, with the sum of coefficients equal to the 12th Fibonacci number 144, has symmetrical intersection of the interval $(-4, 0)$ in five points, so that the point $x = -2$ is the midpoint between the pair $(-2-\sqrt{3}, -2+\sqrt{3})$ as well between the pair $(-3, -1)$. Another polynomial, $f(x) = x^5 + 9x^4 + 28x^3 + 35x^2 + 15x + 1$, with the sum of coefficients equal to the 11th Fibonacci number 89, has asymmetrical intersection of the interval $(-4, 0)$ in five points, something that, in addition to a visual recognition, can be verified numerically. All such polynomials with coefficients from the rows of the modified Pascal's triangle of Figure 8.10 are called Fibonacci-like polynomials (Abramovich & Leonov, 2019). The use of the word like in the name of the polynomials is due to other types of polynomials associated with Fibonacci numbers; for example, Fibonacci polynomials (Webb & Parberry, 1969) and Catalan's Fibonacci polynomials (Koshy, 2001).

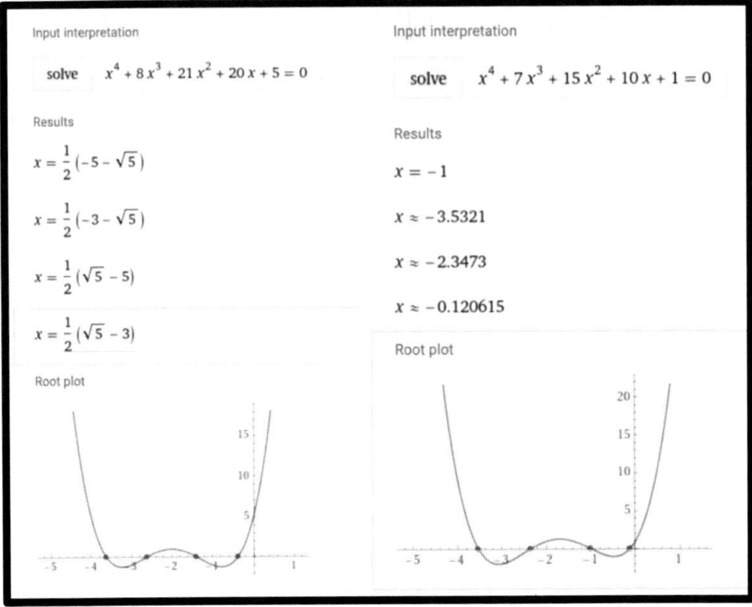

Figure 8.13. Fibonacci-like polynomials of degree four and their roots in Wolfram Alpha.

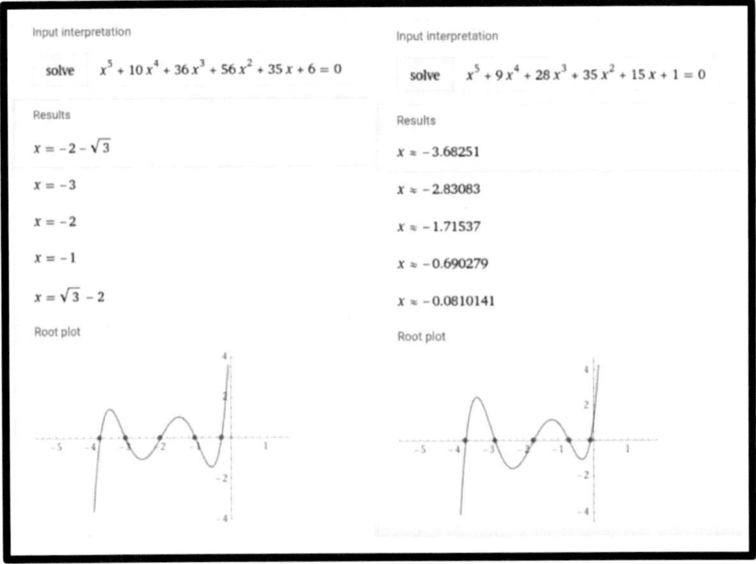

Figure 8.14. Fibonacci-like polynomials of degree five and their roots in Wolfram Alpha.

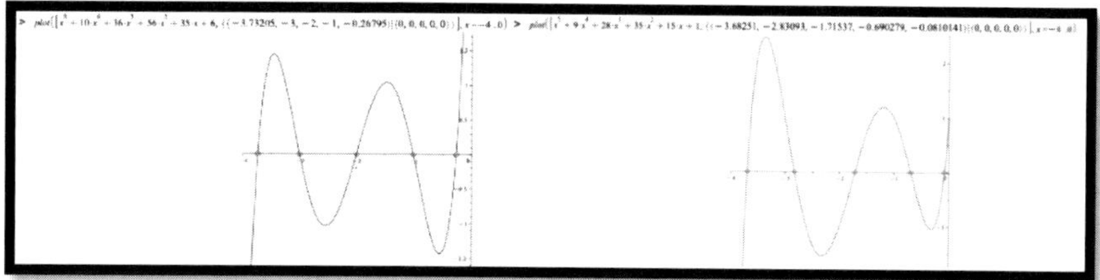

Figure 8.15. Fibonacci-like polynomials of degree five and their roots in Maple.

One can note that within each degree, one of Fibonacci-like polynomials have the unity as its free term. Beginning from the third degree, the polynomials alternate locations of their roots between symmetry and asymmetry, and those having unity as the free term display asymmetry of the roots. Of course, these observations are purely computational.

8.5. Difference Equations and Asymmetry of Generalized Golden Ratios

The roots of Fibonacci-like polynomials, in addition to alternating their type of location within the interval (−4, 0), i.e., symmetrical vs. asymmetrical for each degree, are responsible for entrusting the coefficients of the second order linear difference equation

$$f_{n+1} = af_n + bf_{n-1}, f_0 = 1, f_1 = 2, \tag{8.1}$$

defining Fibonacci numbers in the case $a = b = 1$, with relations that provide cyclic behavior of the ratios f_{n+1}/f_n as n increases, and the larger the degree of a Fibonacci-like polynomial, the larger is the length of the corresponding cycle. In that way, the cycles can be referred to as generalized Golden Ratios (Abramovich & Leonov, 2019).

To clarify, let $r_n = f_n/f_{n-1}$. Then it follows from equation (8.1) that

$$r_{n+1} = a + \frac{b}{r_n}, r_1 = 2\,, \tag{8.2}$$

so that

$$r_2 = a + \frac{b}{2}, r_3 = a + \frac{b}{a+\frac{b}{2}}, \ldots, r_8 = a + \cfrac{b}{a+\cfrac{b}{a+\cfrac{b}{a+\cfrac{b}{a+\cfrac{b}{a+\cfrac{b}{a+\frac{b}{2}}}}}}}, r_9 = a + \cfrac{b}{a+\cfrac{b}{a+\cfrac{b}{a+\cfrac{b}{a+\cfrac{b}{a+\cfrac{b}{a+\cfrac{b}{a+\frac{b}{2}}}}}}}}.$$

In order for the iterations r_n to form cycles of length eight, the relation $r_9 = r_1$, involving a continued fraction for r_9, should hold true. In other words, one has to solve the equation $r_9 = 2$ (see the use of Maple in Figure 8.16). Setting $x = \frac{a^2}{b}$ yields the third-degree polynomial $x^3 + 6x^2 + 10x + 4$ which is exactly the Fibonacci-like polynomial with the (symmetrically located) roots shown in Figure 8.11 (left). One of these roots, $x = -2$, is also the root of the Fibonacci-like polynomial $x + 2 = 0$ the coefficients of which belong to the fourth row of the modified Pascal's triangle of Figure 8.10.

$$solve\left(a + \cfrac{b}{a+\cfrac{b}{a+\cfrac{b}{a+\cfrac{b}{a+\cfrac{b}{a+\cfrac{b}{a+\cfrac{b}{a+\frac{b}{2}}}}}}}} = 2, b\right)$$

$$4 - 2a, -\frac{a^2}{2}, \left(-1 + \frac{\sqrt{2}}{2}\right)a^2, \left(-1 - \frac{\sqrt{2}}{2}\right)a^2$$

$$(x+2)\cdot\left(x - \frac{1}{-1+\frac{\sqrt{2}}{2}}\right)\cdot\left(x - \frac{1}{-1-\frac{\sqrt{2}}{2}}\right)$$

$$(x+2)\left(x - \frac{1}{-1+\frac{\sqrt{2}}{2}}\right)\left(x - \frac{1}{-1-\frac{\sqrt{2}}{2}}\right)$$

$expand(\%)$

$$x^3 + 6x^2 + 10x + 4$$

Figure 8.16. From continued fraction defining cycle to Fibonacci-like polynomial.

Selecting its largest root, $x = \sqrt{2} - 2$, one can iterate recursive relation (8.2) with $a = 1$ within a spreadsheet (Figure 8.17, row 1) to have a (proper) cycle of length eight. Likewise, selecting the smallest root, $x = -\sqrt{2} - 2$, one can iterate recursive relation (8.2) with $a = 1$ within a spreadsheet (Figure 8.17, row 5) to have a (proper) cycle of length eight. However, selecting the root $x = -2$ about which the above two roots are symmetrical, and iterating recursive relation (8.2) with $a = 1$ within a spreadsheet (Figure 8.17, row 9) yields a (proper) cycle of length four which is also a trivial cycle of length eight.

Alternatively, one can take the smallest root, $\frac{a^2}{b} = x \cong -3.2469796037$, out of three asymmetrically located roots of the Fibonacci-like polynomial $x^3 + 5x^2 + 6x + 1$ and iterate recursive relation (8.2) with $a = 1$ within a spreadsheet to have a proper cycle of length seven (Figure 8.17, row 13). Likewise, iterating recursive relation (8.2) using the other two roots of the last polynomial also yields proper cycles of length seven (Figure 8.17, rows 17 and 21). That is, Fibonacci-like polynomials have dual origin, one grounded in their coefficients located in the rows of the modified Pascal's triangle (Figure 8.10) and another grounded in defining cycles formed by solutions of non-linear recursive equation (8.2) with parameters a and b through solving equations with continued fractions and forming polynomials about the ratios $\frac{a^2}{b}$. As it turned out, Fibonacci-like polynomials, stemming from the latter context, have their roots entrusting solutions of recursive relation (8.2) with cycles of different lengths that serve as generalized Golden Ratios.

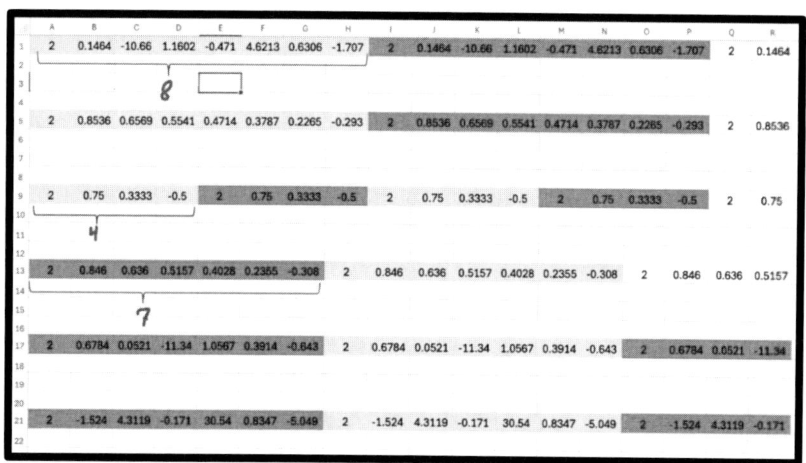

	A	B	C	D	E	F	G	H	I	J	K	L	M	N	O	P	Q	R
1	2	0.1464	-10.66	1.1602	-0.471	4.6213	0.6306	-1.707	2	0.1464	-10.66	1.1602	-0.471	4.6213	0.6306	-1.707	2	0.1464
5	2	0.8536	0.6569	0.5541	0.4714	0.3787	0.2265	-0.293	2	0.8536	0.6569	0.5541	0.4714	0.3787	0.2265	-0.293	2	0.8536
9	2	0.75	0.3333	-0.5	2	0.75	0.3333	-0.5	2	0.75	0.3333	-0.5	2	0.75	0.3333	-0.5	2	0.75
13	2	0.846	0.636	0.5157	0.4028	0.2355	-0.308	2	0.846	0.636	0.5157	0.4028	0.2355	-0.308	2	0.846	0.636	0.5157
17	2	0.6784	0.0521	-11.34	1.0567	0.3914	-0.643	2	0.6784	0.0521	-11.34	1.0567	0.3914	-0.643	2	0.6784	0.0521	-11.34
21	2	-1.524	4.3119	-0.171	30.54	0.8347	-5.049	2	-1.524	4.3119	-0.171	30.54	0.8347	-5.049	2	-1.524	4.3119	-0.171

Figure 8.17. Using a spreadsheet to demonstrate proper and trivial cycles.

In general, asymmetrically located within the interval (−4, 0) roots of Fibonacci-like polynomials generate only proper cycles of prime number length, and symmetrically located within the interval (−4, 0) roots of Fibonacci-like polynomials generate both trivial and proper cycles of even length. However, cycles of odd length can also be comprised of cycles the length of which is a divisor of an odd number. For example, the Fibonacci-like polynomial $x^4 + 7x^3 + 15x^2 + 10x + 1$ has four roots (Figure 8.13, right), three of which generate proper cycles of length nine and one root, $x = -1$ (being the root of the Fibonacci-like polynomial $x + 1 = 0$ the coefficients of which belong to the third row of the modified Pascal's triangle of Figure 8.10), generates a trivial nine cycle comprised of three cycles of length three. To connect cycles to asymmetry, note than a proper cycle of any length represents a string of numbers that cannot be split into like strings of equal numbers to demonstrate symmetry. That is, all proper cycles, by definition, reveal asymmetry. A cycle the length of which is not a prime number may consist of several cycles that are proper cycles of a smaller length. For example, as was mentioned above, a string of three cycles of length three represents a trivial cycle of length nine which is symmetric. At the same time, all cycles of length seven are asymmetric. Computational triangulation using Wolfram Alpha, Maple, and a spreadsheet represents a powerful approach to demonstrating multiple ideas – iteration, symmetry, asymmetry, cycles, and generalized Golden Ratios in the form of the strings of numbers – to the learners of mathematics.

Conclusion

The chapter illustrated the joint use of computational triangulation and the TITE framework when exploring two problems from discrete mathematics. The first, unquestionably famous problem, integrating the theory of numbers and probability theory, demonstrated the use of digital tools in finding the probability of irreducibility of a fraction with randomly selected numerator and denominator from the set of natural numbers. The second problem, stemming from the rearrangement of entries of Pascal's triangle led to Fibonacci-like polynomials the roots of which alternate symmetrical and asymmetrical location within the interval to which they all belong, whatever the degree of a polynomial. Computational triangulation using multiple digital tools made it possible to introduce generalized Golden Ratios in the form of the strings of numbers of different lengths. The scope of this last chapter is appropriate to be included in a course for prospective secondary mathematics teachers.

References

Abramovich, S. (2014). Revisiting mathematical problem solving and posing in the digital era: Toward pedagogically sound uses of modern technology. *International Journal of Mathematical Education in Science and Technology*, *45*(7), 1034-1052.

Abramovich, S. (2015). Mathematical problem posing as a link between algorithmic thinking and conceptual knowledge. *The Teaching of Mathematics, 18*(2), 45–60.

Abramovich, S. (2021). Using Wolfram Alpha with elementary teacher candidates: From more than one correct answer to more than one correct solution. *MDPI Mathematics* (Special issue: Research on Teaching and Learning Mathematics in Early Years and Teacher Training), *9*(17), 2112, 18 pages.

Abramovich, S. (2022). Advancing the concept of triangulation from the social sciences research to mathematics education. *Advances in Educational Research and Evaluation*, *3*(1), 201–217. DOI 10.25082/AERE.2022.01.002

Abramovich, S. (2023). Computational triangulation in mathematics teacher education. *MDPI Computation* (special issue "Computational Social Science and Complex Systems"), *11*(2), 31.

Abramovich, S. (2024). Conceptual shortcuts and technology in mathematical problem solving. In M. G. Voskoglou & J. Stiles (Eds.), *Focus on mathematics education research* (pp. 49–86). Nova Science Publishers.

Abramovich, S. (2025a). *From counting to computing: Ideas for mathematics rducation in information age*. Emerald Publishing.

Abramovich, S. (2025b). From symmetry to asymmetry with digital tools in mathematics teacher education. *Asymmetry* 2025(1): 0002, https://doi.org/10.55092/asymmetry 20250002.

Abramovich, S., & Freiman, V. (2023). *Fostering collateral creativity in school mathematics: Paying attention to students' emerging ideas in the age of technology*. Springer.

Abramovich, S., & Griffin, L. (2025). Cookies on plates: Extending Fibonacci-like numbers to fractions in a 3rd grade classroom. *Forum for Education Studies*, *3*(1), 2310. https://doi.org/10.59400/fes2310.

Abramovich, S., & Leonov, G. A. (2019). *Revisiting Fibonacci numbers through a computational experiment.* Nova Science Publishers.

Abramovich, S., & Nikitin, Ya. Yu. (2017). Teaching classic probability problems with modern digital tools. *Computers in the Schools*, *34*(4), 318–336.

Advisory Committee on Mathematics Education. (2007). Mathematical Needs of 14–19 Pathways. The Royal Society.

Andrews, G. E. ΕΥΡΗΚΑ! num = Δ + Δ + Δ. (1986). *Journal of Number Theory, 23*(3), 285–293.

Archimedes. (1912). *The method of Archimedes*. Heath, T. L. (Ed.) Cambridge University Press.

Arnheim, R. (1969). *Visual thinking*. University of California Press.

Arnold, V. I. (2003). *New obscurantism and Russian education* (in Russian). Fazis. Available at: http://www.mccme.ru/edu/viarn/obscur.htm.

Arnold, V. I. (2015). *Lectures and problems: A gift to young mathematicians*. American Mathematical Society.

Association of Mathematics Teacher Educators. (2017). *Standards for preparing teachers of mathematics*. https://amte.net/standards, accessed on January 19, 2022.

Avitzur, R. (2011). *Graphing calculator* [Version 4.0]. Pacific Tech.

Bachman, H. J., Elliott, L., Duong, S., Betancur, L., Navarro, M. G. Votruba-Drzal, E., & Libertus, M. (2020). Triangulating multi-method assessments of parental support for early math skills. *Frontiers in Education*, 5:589514; doi: 10.3389/feduc.2020.589514.

Bailey, D. H., & Borwein, J. M. (2018). Computation and experimental evaluation of Mordell-Tornheim-Witten sum derivatives. *Experimental Mathematics*, *27*(3), 370–376.

Beiler, A. H. (1966). *Recreations in the theory of numbers*. Dover.

Bell, J., Hare, K., & Shallit, J. (2018). When is an automatic set an additive basis? *Proceedings of American Mathematical Society, Series B 5*, 50–63.

Bernardi, C., & Menghini, M. (2024). The training of mathematics teachers: a higher standpoint. In A. Verdugo Rohrer & J. Zender (Eds.), *History of mathematics and its contexts, trends in the history of science* (pp. 31-43). Springer.

Bilalić, M., McLeod, P., & Gobet, F. (2010). The mechanism of the Einstellung (set) effect: A pervasive source of cognitive bias. *Current Directions in Psychological Science*, *19*(2), 111–115.

Boyer, C. B., & Merzabach, U. C. (1989). *A history of mathematics* (2nd edition). Wiley.

Bowen, M. (1994). *Family therapy in clinical practice*. Jason Aranson.

Cai, J., & Leikin, R. (Eds.) (2025). *Research in mathematical problem posing: New advances and directions*. Springer.

Campbell, D. T., & Fiske, D. W. (1959). Convergent and discriminant validation by the multitrait-multimethod matrix. *Psychological Bulletin, 56* (2), 81–105.

Canobi, K. H. (2005). Children's profiles of addition and subtraction understanding, *Journal of Experimental Child Psychology*, *92*(3), 220–246.

Catalan, E. (1884). Notes sur la théorie des fractions continues et sur certaines séries. *Mémoires de L'Académie Royale des Sciences, des Lettres et des Beaux-Arts de Belgique*, Tome XLV. Bruxelles: F. Hayez, Imprimeur de L'Académie Royale.

Cevizci, B. (2018). How and why does the multiplication method developed by the Russian peasants work? *Journal of Inquiry Based Activities*, *8*(1), 24–36.

Chace, A. B., Manning, H. P., & Archibald, R. C. (1927). *The Rhind Mathematical Papyrus, British museum 10057 and 10058, vol. 1*. Mathematical Association of America.

Char, B. W., Geddes, K. O., Gonnet, G. H., Leong, B. L., Monogan, M. B., & Watt, S. M. (1991). *Maple V language reference manual*. Springer.

Cho, S., & Kim, J. (2013). Effects of learner-centered instruction on learners' reasoning ability: focus on third-grade division. In J. Kim, I. Han, & M. K, J. Lee (Eds.), *Mathematics education in Korea. Volume 1: Curricular and teaching and learning practices* (pp. 130–152). World Scientific.

Common Core State Standards. (2010). *Common core standards initiative: Preparing America's students for college and career.* http://www.corestandards.org, accessed on October 29, 2025.

Conference Board of the Mathematical Sciences. (2012). *The mathematical education of teachers II.* The Mathematical Association of America.

Conole, G., & Dyke, M. (2004). What are the affordances of information and communication technologies? *ALT-J: Research in Learning Technology, 12*(3), 113–124.

Cooney, T. J. (2001). Considering the paradoxes, perils, and purposes of conceptualizing teacher development. In F. L. Lin (Ed.), *Making sense of mathematics teacher education* (pp. 9–31). Kluwer.

Denzin, N. K. (1970). *The research act in sociology: The theoretical introduction to sociological methods*. Butterworth.

Denzin, N. K. (2007). Triangulation. In G. Ritzer (Ed.), *The Blackwell encyclopedia of sociology*, vol. X (pp. 5075–5080). Blackwell Publishing.

Department for Education (2013, updated 2021). *National curriculum in England: Mathematics programmes of study*. Crown copyright. https://www.gov.uk/government/publications/national-curriculum-in-england-mathematics-programmes-of-study, accessed on October 29, 2025.

Department of Basic Education. (2018). *Mathematics teaching and learning framework for South Africa: Teaching mathematics for understanding*. The Author.

Dunker, K. (1945). On problem solving. *Psychological Monographs*, 58 (whole No. 270).

Elkies, N. D. (1988). On $A^4 + B^4 + C^4 = D^4$. *Mathematics of Computation, 51*(184), 825–835.

Felmer, P., Lewin, R., Martínez, S., Reyes, C., Varas, L., Chandía, E., Dartnell, P., López, A., Martínez, C., Mena, A., Ortíz, A., Schwarze, G., & Zanocco, P. (2014). *Primary mathematics standards for pre-service teachers in Chile*. World Scientific.

Freudenthal, H. (1978). *Weeding and sowing*. Kluwer.

Gauss, C. F. (1966). *Disquisitiones arithmeticae* (Arithmetic disquisitions; trans. from Latin by A. A. Clarke). Yale University Press.

Gimmestad, B. J. (1991). The Russian peasant multiplication algorithm: A generalization. *The Mathematical Gazette, 75*(472), 169–171.

Grinshpan, A. Z. (1999). The Bieberbach conjecture and Milin's functionals. *American Mathematical Monthly, 106*(3), 203–214.

Harrison, J. (2008). Formal proof – theory and practice. *Notices of American Mathematical Society, 55*(11), 1395–1406.

Helfgott, H. A. (2014). The Ternary Goldbach Conjecture is true. *arXiv:1312.7748v2.* https://doi.org/10.48550/arXiv.1312.77483.

Hill, D. R. (2000). Engineering. In R. Rashed (Ed.) *Encyclopedia of the history of Arabic science,* vol 3 (pp.751-795). Routledge.

Hohenwarter, M. (2002). *GeoGebra.* [Computer software]. Available at: https://www.geogebra.org/download. Accessed on October 29, 2025.

Isoda, M. (Ed.) (2010). Junior high school teaching guide for the Japanese course of study: Mathematics. Center for Research on International Cooperation in Educational Development (CRICED), University of Tsukuba. [Online: www.criced.tsukuba.ac.jp/math/apec/ICME12/Lesson_Study_set/.

JuniorHighSchoolTeaching Guide-Mathmatics-JP-EN.pdf].

Kafoussi, S., & Margaritidou, C. (2023). Integrating the history of mathematics in mathematics education: Examples and reflections from the classroom. In S. Romero Sanchez, A. Serradó Bayés, P. Appelbaum, & G. Aldon (Eds.), *The role of the history of mathematics in the teaching/learning process. Advances in mathematics education.* Springer, https://doi.org/10.1007/978-3-031-29900-1_6.

Kitcher, P. (1983). *The nature of mathematical knowledge*. Oxford University Press.

Klein, F. (2016). *Elementary mathematics from a higher standpoint. Volume 1: Arithmetic, algebra, analysis*. Springer-Verlag.

Kline, M. (1985). *Mathematics for the non-mathematician.* Dover.

Koshy, T. (2001). *Fibonacci and Lucas numbers with applications*. John Wiley & Sons.

Koshy, T. (2014). *Pell and Pell-Lucas numbers with applications.* Springer.

Kuijt, I. (Ed.) (2002). *Life in neolithic farming communities: Social organization, identity and differentiation*. Kluwer.

Kuo, E., Hull, M. M., Gupta, A., & Elby, A. (2013). How students blend conceptual and formal mathematical reasoning in solving physics problems. *Science Education*, *97*(1), 32–57.

Lander, L. J., & Parkin, T. R. (1966). Counterexample to Euler's conjecture on sums of like powers. *Bulletin of the American Mathematical Society*, *72*(6), 1079.

Langtangen, H. P., & Tveito, A. (2001). How should we prepare the students of science and technology for a life in the computer age? In B. Engquist & W. Schmid (Eds.), *Mathematics unlimited—2001 and beyond* (pp. 809-825). Springer.

Laplace, P. S. (1814/1951). *A philosophical essay on probabilities*. Dover.

Leonov, G. A., & Kuznetsov, N. V. (2013). Hidden attractors in dynamical systems. From hidden oscillations in Hilbert–Kolmogorov, Aizerman, and Kalman problems to hidden chaotic attractors in Chua circuits. *International Journal of Bifurcation and Chaos 23*(01), 1330002-1–69.

Levitt, E. E. (1956). The water-jar Einstellung test as a measure of rigidity. *Psychological Bulletin*, *53*, 347–370.

Liljedahl, P., Oesterle, S., & Bernèche, C. (2012). Stability of beliefs in mathematics education: A critical analysis. *Nordic Studies in Mathematics Education, 17*(3-4), 101–118.

Löwe, B., & van Kerkhove, B. (2019). Methodological triangulation in empirical philosophy (of mathematics). In A. Aberdein & M. Inglis (Eds.), *Advances in experimental philosophy of logic and mathematics* (pp. 15–38). Bloomsbury.

Luchins, A. S. (1942). Mechanization in problem solving: The effect of Einstellung. *Psychological Monographs*, *54*(6), whole No. 248. The American Psychological Association.

Lyness, R. C. (1968). Applied mathematics in English schools. *Educational Studies in Mathematics, 1*(1/2), 98-104.

Mason, J., & Pimm, D. (1984). Seeing the general in the particular. *Educational Studies in Mathematics, 15*(3), 277–289.

Marx, K. (1982). *Das Kapital*, vol. 1, Part 3, Chapter 5. Penguin Books.

McFee, G. (1992). Triangulation in research: Two confusions. *Educational Research, 34*(3), 215–219.

Metropolis, N., & Ulam, S. (1949). The Monte Carlo method. *Journal of American Statistical Association, 44*(247), 335–341.

Ministry of Education Singapore. (2012). *Mathematics syllabus, primary one to six.* The Author. https://www.moe.gov.sg//media/files/primary/mathematics_syllabus_primary_1_to_6.pdf. Accessed on October 29, 2025.

Ministry of Education Singapore. (2020). *Mathematics syllabuses, secondary one to four*, The Author. https://www.moe.gov.sg/-/media/files/secondary/syllabuses/maths/2020-express_na-maths_syllabuses.pdf?la=en&hash=95B771908EE3D777F87C5D6560EBE6DDAF31D7EF. Accessed on October 29, 2025.

Mironov, B. N. (2019). The cognitive abilities of the Russian peasantry at the turn of the twentieth century. *Russian History, 46*(2-3), 103–124.

Mok, I. A. C., & Clarke, D. (2015). The contemporary importance of triangulation methods in post-positivist world. In A. Bikner-Ahsbahs, C. Knipping, & N. Presmeg (Eds.), *Approaches to qualitative research in mathematics education* (pp. 403–425). Springer.

Nathanson, M. B. (1987). A short proof of Cauchy's polygonal number theorem. *Proceedings of American Mathematical Society, 99*(1), 22–24.

National Council of Teachers of Mathematics. (2000). *Principles and standards for school mathematics*. The Author.

National Council of Teachers of Mathematics. (2014). *Principles to actions: Ensuring mathematical success for all.* The Author.

National Curriculum Board. (2008). *National mathematics curriculum: Framing paper.* The Author. http://www.acara.edu.au/verve/_resources/National_Mathematics_Curriculum_-_Framing_Paper.pdf. Accessed on October 29, 2025.

Ontario Ministry of Education. (2020). *The Ontario curriculum, grades 1–8, mathematics (2020)*, http://www.edu.gov.on.ca. Accessed on October 29, 2025.

Otun, I. W., & Njoku, O. G. (2020). Developing pre-service mathematics teachers' mathematical problem solving-posing skills through solve-reflect-pose strategy in Lagos state, Nigeria. *Journal of Educational Research in Developing Areas, 1* (2), 140–152. https://doi.org/10.47434/JEREDA/1.2.2020.140.

Park, M., & Choi-Koh, S. (2013). Future directions for mathematics textbooks. In J. Kim, I. Han, & J. K. Lee (Eds.), *Mathematics education in Korea. Volume 1: Curricular and teaching and learning practices* (pp. 80–103). World Scientific.

Pólya, G. (1954). *Mathematics and plausible reasoning. Volume 1: Induction and analogy in mathematics*. Princeton University Press.

Pólya, G. (1973). *How to solve it.* Princeton University Press.

Rohlin, V. A. (2013). A lecture about teaching mathematics to non-mathematicians. Available at https://www.math.stonybrook.edu/~oleg/Rokhlin/LectLMO-eng.html/.

Russell, B. (1945.) *A history of Western philosophy*. Simon and Schuster.

Sharma, S. (2013). Qualitative approaches in mathematics education research: Challenges and possible solutions. *Education Journal*, *2*(2), 50–57.

Swan, M. (2007). The impact of task-based professional development on teachers' practices and beliefs: A design research study. *Journal of Mathematics Teacher Education, 10*(4), 217–237.

Tikhomirov, V. M. (2001). A. N. Kolmogorov. In S. S. Chern & F. Hirzebruch (Eds.), *Wolf Prize in mathematics*, vol. 2 (pp. 119–164). World Scientific.

Tzanakis, C., & Arcavi A. (2002). Integrating history of mathematics in the classroom: an analytic survey. In J. Fauvel & J. van Maanen (Eds.), *History in mathematics education. The ICMI study* (pp. 201–240). Kluwer.

Van Bendegem, J. P. (1998). What, if anything, is an experiment in mathematics? In D. Amapolitanos, A. Baltas, & S. Tsinorema (Eds.), *Philosophy and the many faces of science* (pp. 172–182). Rowman & Littlefield.

Vavilov, N. A. (2021). Computers as novel mathematical reality. IV. Goldbach Problem. *Computer Tools in Education*, *4*, 5–71.

Von Mises. R. (1957). *Probability, statistics and truth*. MacMillan.

Vygotsky, L. S. (1930). The instrumental method in psychology (talk given in 1930 at the Krupskaya Academy of Communist Education). *Lev Vygotsky archive* [Online materials]. Available at: https://www.marxists.org/archive/vygotsky/works/1930/instrumental.htm. Accessed on October 29, 2025.

Vygotsky, L. S. (1999). Consciousness as a problem in the psychology of behavior. Undiscovered Vygotsky: Etudes on the pre-history of cultural-historical psychology. *European Studies in the History of Science and Ideas*, vol. 3, pp. 251-281. Peter Lang Publishing.

Webb, E. J., Campbell, D. T., Schwartz, R. D., & Sechrest, L. (1966). *Unobtrusive measures*. Rand McNally.

Webb W., & Parberry, E. (1969). Divisibility properties of Fibonacci polynomials. *Fibonacci Quarterly*, *7*(5), 457–463.

Wertheimer, M. (1959). *Productive thinking*. Harper & Row.

Western and Northern Canadian Protocol. (2008). *The Common Curriculum Framework for Grades 10-12 Mathematics* [on-line materials]. Available at: http://www.bced.gov.bc.ca/irp/pdfs/mathematics/WNCPmath1012/2008math1012wncp_ccf.pdf. Accessed on October 29, 2025.

Wing, J. M. (2006). Computational thinking. *Communications of the ACM*, *49*(3), 33–35.

Index

L

M

N

O

P

Q

R

S

T

U